U0281058

水利

造福民生的伟大事业

Water Conservancy

A Great Cause for People's Livelihood in China

中华人民共和国水利部　编

Compiled by the Ministry of Water Resources, P.R. China

造福民生永无止境。当前，我国已进入全面建成小康社会的决定性阶段。党的十八大把水利放在生态文明建设的突出位置，对水利工作作出重要部署，提出更高要求。我们必须清醒地认识到人多、地少、水缺仍然是我国的基本国情和水情，洪涝灾害、干旱缺水、水体污染、水土流失问题已成为制约经济社会可持续发展的主要瓶颈。治水兴水是一项复杂的系统工程和长期的艰巨任务，我们必须从保障国家安全的高度认识水资源问题，立足经济社会全局看待水利工作，牢固树立以人为本的发展理念，全力倡导服务人民的价值取向，始终确立民生为先的实践要求，奋力实现造福民生的目标追求。

『潮平两岸阔，风正一帆悬。』站在新的历史起点上，我们要乘着党的十八大东风，以科学发展观为指导，深入贯彻落实中央兴水惠民决策部署，不断谱写民生水利发展新篇章，让江河更加安澜、山川更加秀美、人民更加安康，让水利更好地造福中华民族、惠泽子孙后代。

是为序。

中华人民共和国水利部部长

序

水是生命之源、生产之要、生态之基。

水利是造福民生的伟大事业。

由于特殊的地理和气候条件，我国始终面临着非同寻常的艰巨治水任务。水利不仅为人民生存发展、安居乐业所必需，也历来是治国安邦的大事。圣人之治，其枢在水，是历朝历代治理天下的成功经验；以人为本、造福民生，则是新时期水利工作的本质要义和实践追求。

大力发展民生水利，是科学发展观在水利领域的重要体现，也是水利人的庄严承诺和生动实践。近年来，各级水利部门集中力量办了很多事关国计民生的大事、关系群众切身利益的好事，推动民生水利发展取得显著成效，受到人民群众的拥护和赞誉。在应对洪涝干旱灾害过程中，我们始终坚持以人为本、生命至上的理念，最大程度保障受灾群众的生命财产安全和饮水安全；在推进水利建设过程中，我们始终坚持把人民群众直接受益的基础设施作为优先领域，干成了一批百姓看得见、摸得着、得实惠的惠民工程；在加强水利管理过程中，我们始终注重妥善处理好各种利益关系，切实保障人民群众在水资源开发利用、城乡供水保障、水利移民安置等方面的合法权益；在深化水利改革的过程中，我们始终注重倾听和接受广大基层群众的意见，以是否更有利于实现好、保障好、维护好人民群众的利益为标准，判断水利改革的成败得失。水利部编辑出版这本画册，旨在真实记录新时期水利跨越发展的新进程，生动展现新时期民生水利实践探索的新成就。

Preface

Water is the source of life, the necessity of production, and the basis of ecology.

Water conservancy is a great cause benefitting people's livelihood.

China is always confronted with the exceptionally difficult task of water governance due to her special geographical and climatic conditions. Water conservancy is not only a necessity for people's development and livelihood, but also a crucial factor for the prosperity and stability of the country. A lesson learned through various periods of China's history is best described as "A sage takes water governance as the core of his rule". In modern times, putting people first and creating benefits for the people is an essential principle of water conservancy work and a practical pursuit of water professionals.

To vigorously develop water conservancy for people's livelihood is an important manifestation of the Scientific Outlook on Development in the water sector, as well as a solemn promise made and a practical action taken by water professionals. In recent years, water departments at all levels have made remarkable progress in the development of water conservancy for people's livelihood by fulfilling a number of major tasks concerning national interests and people's livelihood, and many good endeavors closely related to people's benefits, winning strong support and high praise from the people. To cope with floods and droughts, we have insisted on the principle of putting people's life as the top priority, making the best efforts to protect people's lives, property and drinking water safety in disaster-stricken areas. In the process of water sector development, we have placed great emphasis on infrastructure projects that create the most direct benefits for the people, completing a series of projects that are visible, accessible and beneficial to the people. Throughout the process of enhancing water management, we have persisted in our effort to properly handle the relationship among various stakeholders, effectively safeguarding people's rights concerning water resources exploitation and utilization, urban and rural water supply, resettlement of project affected communities, ect. In furthering water sector reform, we listen to the opinions of the public and respect public views and suggestions. Whether or not a proposed reform initiative could help realize, guarantee and safeguard people's interests has been used as the criteria to evaluate its success and effectiveness. By publishing this album, the Ministry of Water Resources of China intends to record the progress of water conservancy development in the new era. This album also aims at presenting China's latest achievements in practicing and exploring water conservancy for

people's livelihood.

Bringing benefits to the people knows no end. At present, China has entered a crucial stage in completing the building of a moderately prosperous society in all respects. The 18th CPC National Congress placed ecological civilization development at a prominent position, making key deployment and raising higher requirements for water conservancy work. We must be aware that a large population, insufficient land resources and the shortage of water resources still constitute the basic national situation and water conditions of China. Floods, droughts, water pollution and water and soil erosion have become major bottlenecks restricting sustainable economic and social development. Water governance and development is a complex and systematic project as well as a long-term arduous task. We must look at water issues from the perspective of ensuring national water security, consider water conservancy work in the light of the overall situation of socio-economic development, firmly establish the concept of people-oriented development, actively promote the values of serving the people, emphasize the practical requirements of prioritizing people's livelihood, and strive to create more benefits for the people.

"At rising tide the river flows wide, and in fair wind a sail is raised high". Standing at the new starting point, the water sector of China will seize the opportunities brought about by the 18th CPC National Congress, follow the guideline of the Scientific Development Outlook, implement the decisions made by Central Government on developing water resources and benefitting the people, and open new chapters in the history of managing water for people's livelihood, so as to ensure the peace and safety of rivers, the beauty and charisma of mountains, the affluence and safety of the people. Water conservancy will bring more benefits to the Chinese nation and our future generations.

陈雷

Chen Lei

Minister of Water Resources, People's Republic of China

四川九寨沟诺日朗瀑布

Nuorilang Waterfall in Jiuzhaigou Valley, Sichuan Province

目录

浙江桐乡市乌镇河道

Wuzhen Located in Tongxiang City, Zhejiang Province

Contents

贵州荔波县小七孔生态区

The Xiaoqikong Ecological Area of Libo County, Guizhou Province

第一篇

Chapter One

基本国情水情

Basic Conditions of Water Resources in China

我国地理气候条件特殊，人口众多，是世界上水情最复杂、治水任务最繁重、江河治理难度最大的国家。自古以来，中华民族就用智慧和汗水谱写出可歌可泣的治水史，留下了都江堰、灵渠、京杭运河等许多著名的水利工程。新中国成立以来，特别是改革开放以来，党和国家高度重视水利工作，开展了气壮山河的水利建设，在中华大地上绘就了气势恢宏的水利画卷。

With special geographical and climatic conditions, and a large population, China is a country with the most complicated water conditions, the greatest challenges in water regulation, and the most difficulties in harnessing rivers. Ever since ancient times, the Chinese people have written an epic history of water harnessing with their wisdom and sweat, leaving behind a number of world renowned water projects, including Dujiang Weir, Lingqu Canal, and Beijing-Hangzhou Canal. Since the founding of New China, especially since the launch of reform and opening-up, the Chinese government has attached great importance to the water sector, and embarked on an ambitious and impressive water development programme, painting a grandiose picture on the land of China.

我国特殊的地理气候条件

我国位于亚欧大陆东部、太平洋西岸，地势西高东低，呈三级阶梯分布。具有显著的季风气候特征，夏季高温多雨，冬季寒冷少雨，时常出现连续丰水年或连续枯水年，灾害性天气频发重发，其中洪涝、干旱、台风是对我国影响最为严重的自然灾害。

China's Special Geographical and Climatic Conditions

Located in the east of the Eurasian continent, to the west of the Pacific Ocean, the country has a stair-like topography, which is high in the west and low in the east. With a typical monsoon climate, China has high temperature and frequent rainfall in the summer and cold and dry weather in the winter. It is not unusual to have consecutive wet years or dry years and frequent occurrence of disastrous weather conditions. Floods, droughts and typhoons are natural disasters that have the most severe impact on the country.

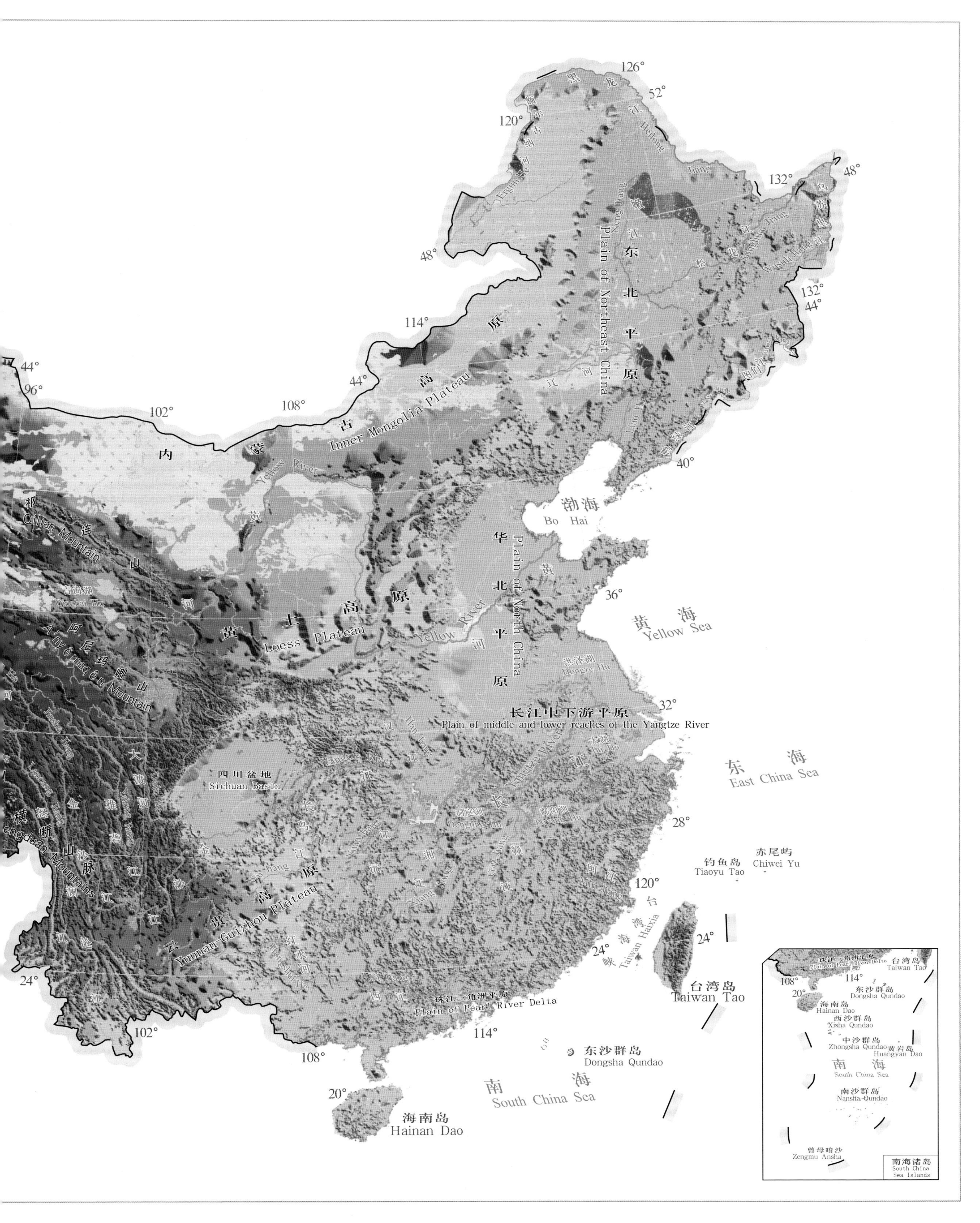

中国地势地貌图

Topographical Map of China

人多水少

我国淡水资源总量为2.84万亿立方米，占全球水资源的6%，排在巴西、俄罗斯、加拿大、美国和印度尼西亚之后，居世界第六位，但人均水资源占有量2 100多立方米，仅为世界平均水平的28%，列世界第125位。

A Large Population with Insufficient Water

China's total volume of freshwater resources amounts to 2.84 trillion m^3, accounting for 6% of the world's total, ranking the sixth in the world, behind Brazil, Russia, Canada, USA and Indonesia, while the per capita water resources stand at 2,100 m^3 only, ranking the 125th in the world and only 28% of the world's average.

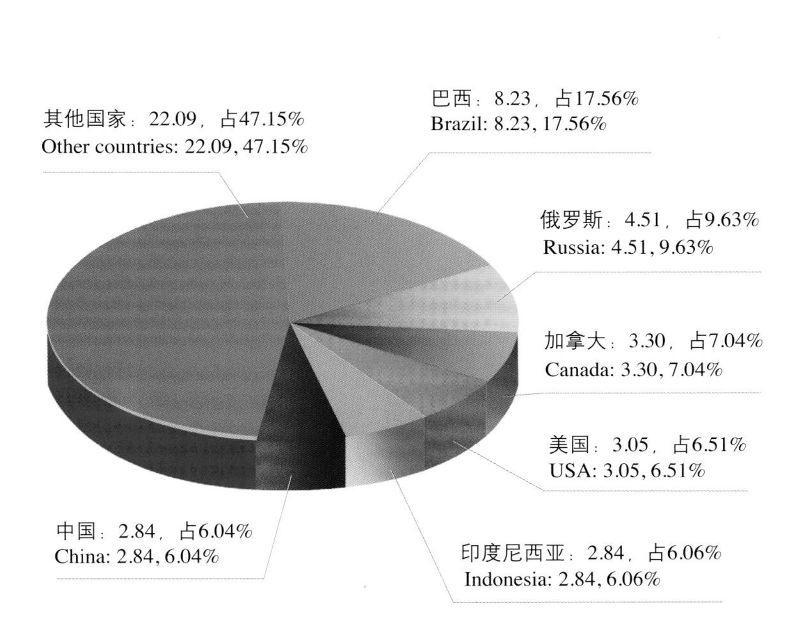

世界部分国家水资源占有量概况

Water resources possession of selected countries

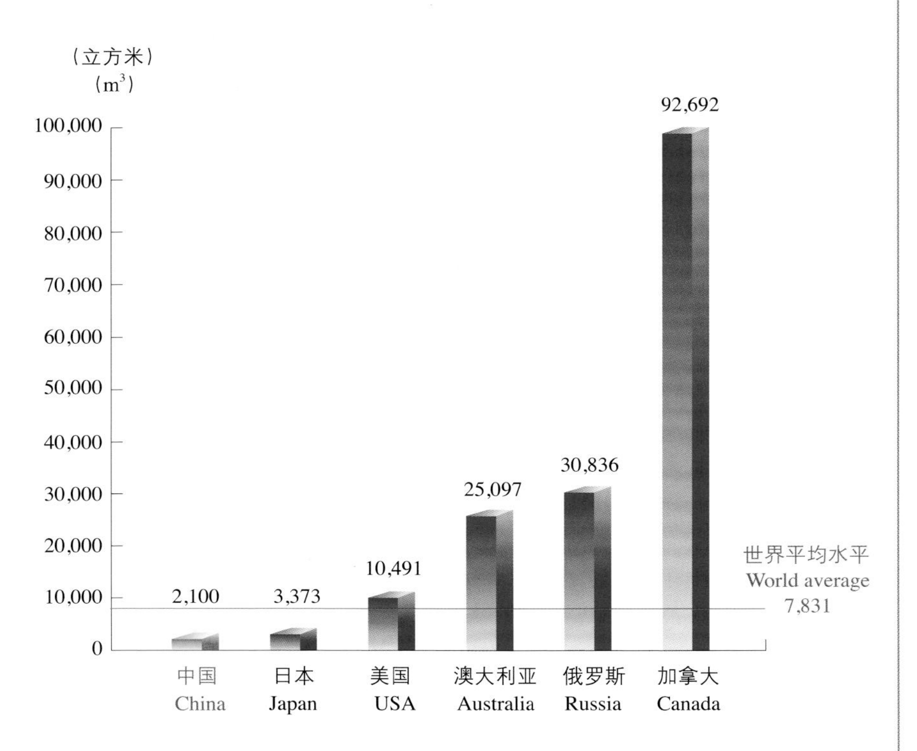

中国人均水资源占有量与世界及部分国家人均水资源占有量对比图

Comparison chart of per capita share of water resources between China and other countries

内蒙古巴丹吉林沙漠

The Badain Jaran Desert, Inner Mongolia Autonomous Region

时空分布不均

我国水资源时空分布不均，年内和年际变化较大，60%～80%集中在汛期，容易形成春旱夏涝以及连涝连旱。水资源与土地、矿产资源分布以及生产力布局也不相匹配，南方多、北方少，北方地区（长江流域以北）面积占全国的63.5%，而水资源仅占全国的19%。

Uneven Distribution of Water in Time and Space

With uneven distribution in time and space, China's water resources vary greatly both within a year and between years. Between 60% and 80% of rainfall is concentrated in the flood season, which easily causes spring droughts and summer deluges, as well as consecutive droughts and floods. The distribution of water resources does not match that of land and mineral resources and productivity. South China is rich in water while North China has insufficient water. Northern China (north of the Yangtze River basin) accounts for 63.5% of China's total territory, while its water resources only account for 19% of the country's total.

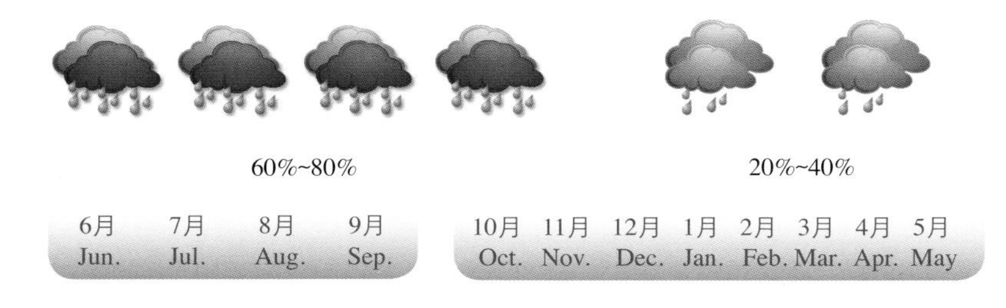

中国年内降水量分布图

Monthly precipitation distribution in China

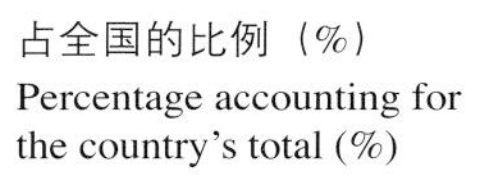

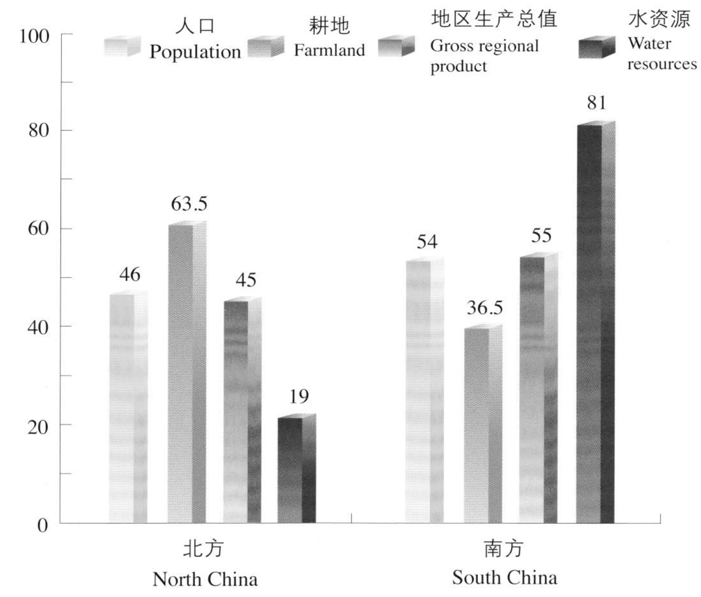

中国人口、耕地、地区生产总值及水资源对比图

Comparison chart of population, farmland, gross regional product and water resources

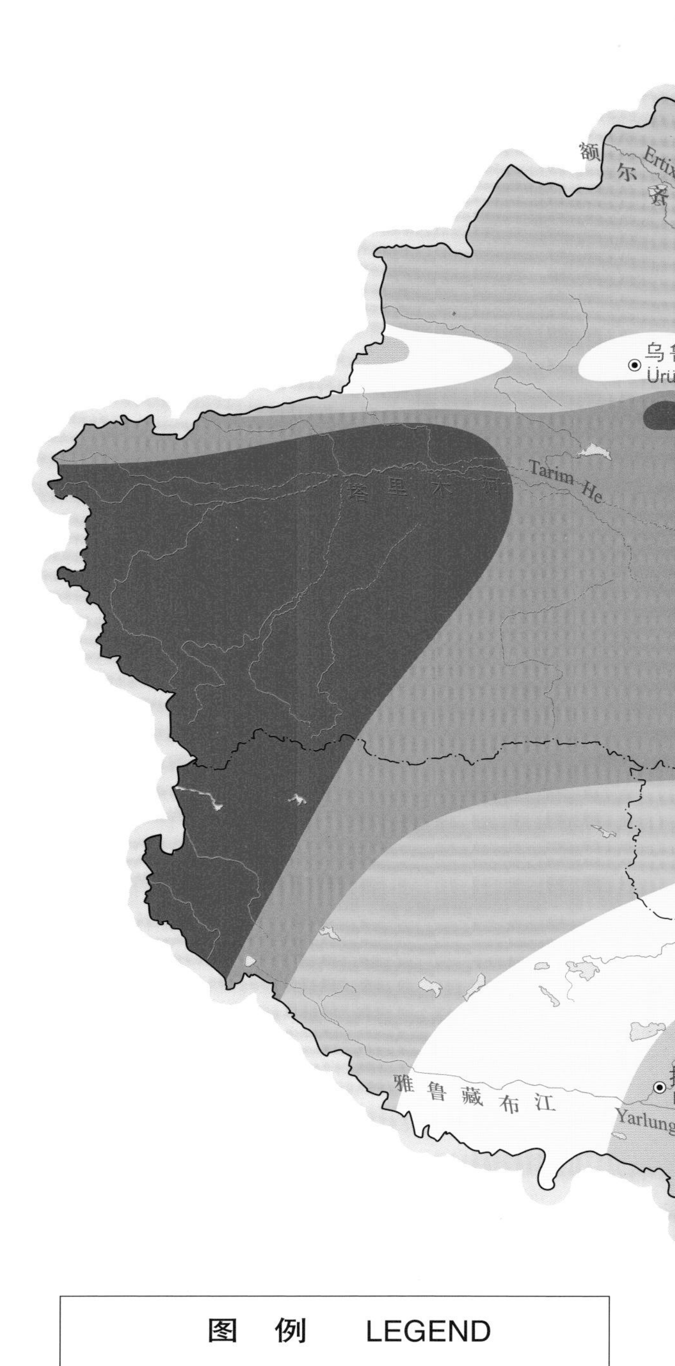

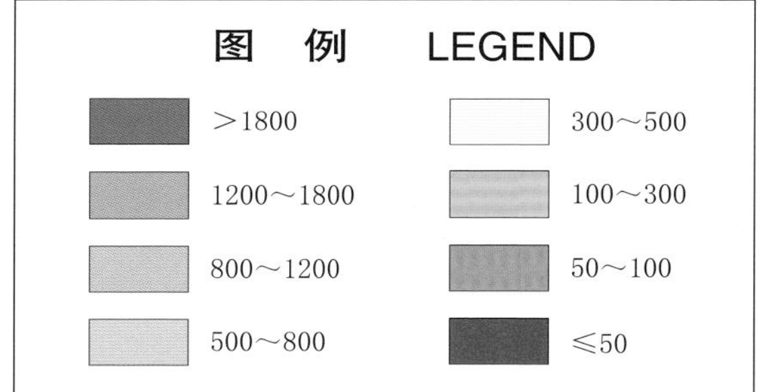

台湾省资料暂缺

Temporary lack of Taiwan Province information

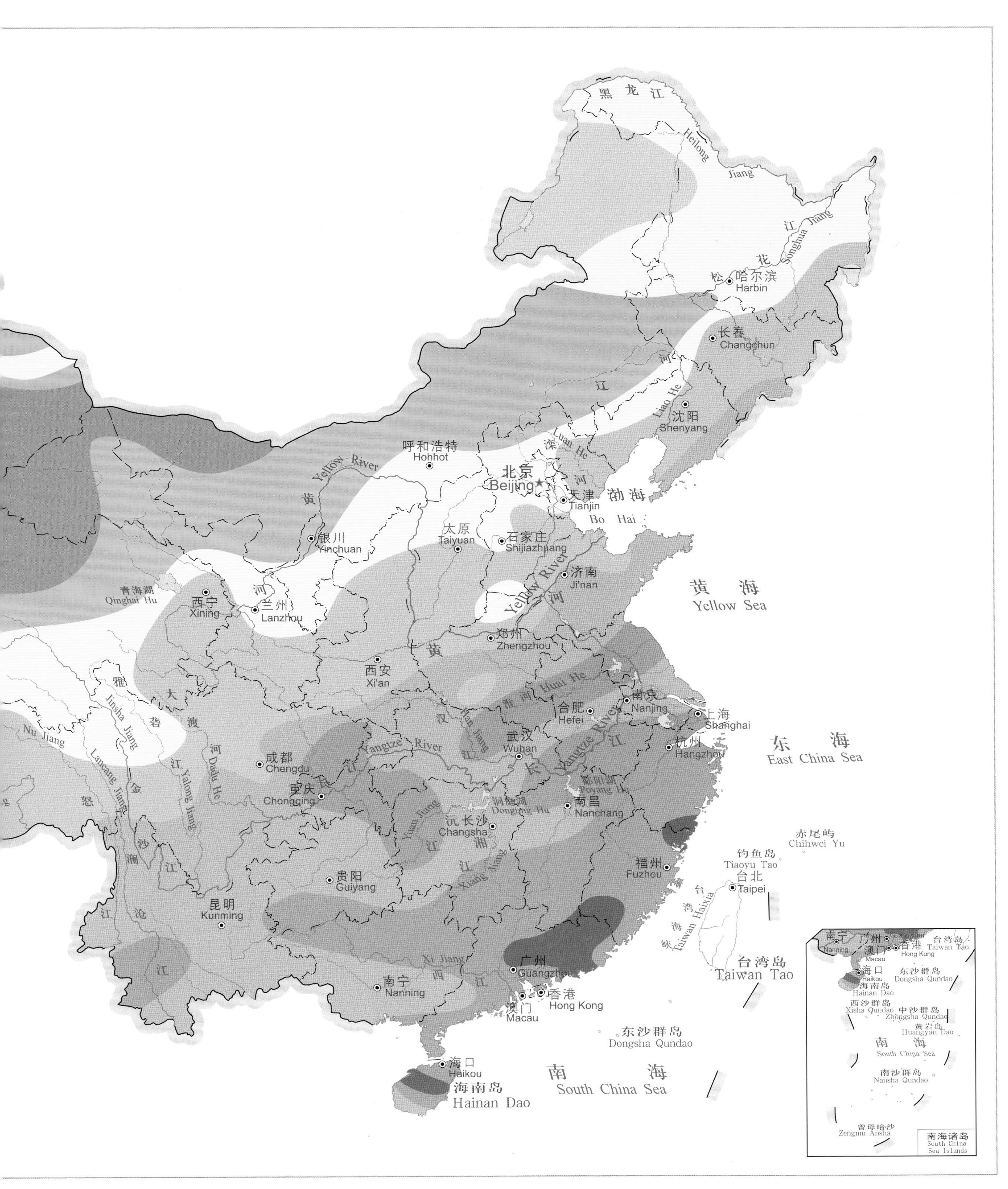

中国年降水量分布图

Annual precipitation distribution in China

四大水问题

Four Major Water Problems

水多——洪涝灾害

1949年以来我国共发生较大洪水50多次，有2/3的国土面积受到各种类型、不同程度洪水的威胁，特别是聚集着全国50%以上人口、35%耕地、2/3工农业总产值的东部和南部经济较发达地区，是遭受洪水威胁最为严重的地区。

Excessive water–floods

Since 1949, China has encountered over 50 major floods, which posed threats at different levels to two-thirds of the nation's territory. In particular, developed regions in the east and south representing over 50% of the country's population, 35% of the farmland and two-thirds of gross agricultural and industrial output, are under the most severe threat of floods.

水少——干旱缺水

我国大部分地区面临不同程度的干旱威胁。1949年以来，发生较大范围的严重干旱17次；2000年以来年均旱灾损失占同期GDP的0.91%；全国660多座城市中，有2/3的城市缺水。

Insufficient water–droughts

Most regions of China are challenged by droughts of different levels. Since 1949, the country has witnessed as many as 17 severe droughts affecting vast areas. Since 2000, the annual economic loss caused by droughts has accounted for 0.91% of the GDP on average. Two-thirds of 660-add cities in China are short of water.

水脏——水体污染

目前我国七大水系、主要湖泊、近岸海域及部分地区的地下水受到不同程度的污染。在调查评价的17.6万公里河长中，Ⅲ类水河长占26.6%，Ⅳ类水河长占13.1%，Ⅴ类水河长占7.8%，劣Ⅴ类水河长占17.7%。

Unclean water–water pollution

China's seven river systems, major lakes, coastal waters and groundwater in some regions suffer from pollution to various degrees. In the 176,000 km of river length investigated, class III water accounts for 26.6%, class IV 13.1%, class V 7.8%, and class V-inferior 17.7%.

水浑——水土流失

我国是世界上水土流失最严重的国家之一，全国水土流失面积356万平方公里，占国土面积的37%，每年流失土壤超过45亿吨，损毁耕地100多万亩。由于面积大、分布广、类型多、治理难、危害重，水土流失已成为我国面临的突出生态和环境问题。

Turbid water–soil erosion

China is one of those countries suffering the most serious soil erosion, with 3.56 million km^2 of soil erosion area, accounting for 37% of the total land area of the country. Annually, 4.5 billion tons of soil is lost to erosion, damaging over 66,000 hectares of farmland.

主要江河湖泊

我国江河湖泊众多。流域面积大于100平方公里的河流有5万多条，流域面积在1 000平方公里以上的河流有1 500多条。最主要的江河有长江、黄河、淮河、海河、珠江、松花江、辽河等。最主要的淡水湖泊有鄱阳湖、洞庭湖、太湖、洪泽湖、巢湖等，最大的咸水湖是青海湖。

Major Rivers and Lakes

China boasts of many rivers and lakes. There are more than 50,000 rivers with a basin area of over 100 km^2, and 1,500 rivers over 1,000 km^2. Major rivers include the Yangtze River, the Yellow River, the Huaihe River, the Haihe River, the Pearl River, the Songhua River, and the Liaohe River. Main freshwater lakes include Poyang Lake, Dongting Lake, Taihu Lake, Hongze Lake, and Chaohu Lake. The biggest saltwater lake is Qinghai Lake.

中国水系图

Major river systems in China

长 江

发源于青藏高原唐古拉山脉主峰各拉丹冬雪山西南侧，干流流经11个省、自治区、直辖市，全长6 397公里，是我国第一、世界第三大河，流域面积180多万平方公里，占全国国土面积的18.8%，多年平均水资源总量9 960亿立方米，约占全国水资源总量的36.5%。

The Yangtze River

The Yangtze River originates on the southwestern side of Geladandong Snow Mountain, the main peak of Tangla Mountains in Qinghai-Tibet Plateau, with the main stream running through 11 provinces, autonomous regions, and municipalities. With a total length of 6,397 km, the Yangtze River is the longest in China and the third-longest in the world. The catchment area covers over 1.8 million km^2, 18.8% of the total land area of the country. The average water volume is 996 billion m^3, accounting for 36.5% of national total.

黄 河

发源于青藏高原巴颜喀拉山北麓，流经9个省、自治区，全长5 464公里，是我国第二、世界第五长河，也是世界上含沙量最多的河流。流域面积75万平方公里，全流域平均年径流量580亿立方米。

The Yellow River

Originating on north Bayan Har Mountains in the Qinghai-Tibet Plateau, the Yellow River runs through nine provinces and autonomous regions. With a total length of 5,464 km, it is the second-longest in China and the fifth-longest in the world, as well as the world's most sedimented river. With a catchment area of 750,000 km^2, the average annual runoff in the whole basin is 58 billion m^3.

淮　河

发源于河南南部桐柏山主峰太白顶，流经湖北、河南、安徽、江苏四省，与秦岭一同组成我国南北方的分界线，流域面积19万平方公里，多年平均水资源量854亿立方米。

The Huaihe River

Originating from Mount Taibai, the main peak of Tongbo Mountain in southern Henan Province, the Huaihe River runs through Hubei, Henan, Anhui, and Jiangsu provinces, forming the geographical dividing line of South and North China together with the Qinling Mountains. The total catchment area is 190,000 km^2, and the average annual ranoff is 85.4 billion m^3.

海　河

系我国华北地区流入渤海诸河的总称，由北运河、永定河、大清河、子牙河、南运河、滦河、徒骇马颊河等组成。流域面积31万多平方公里，多年平均水资源量372亿立方米。

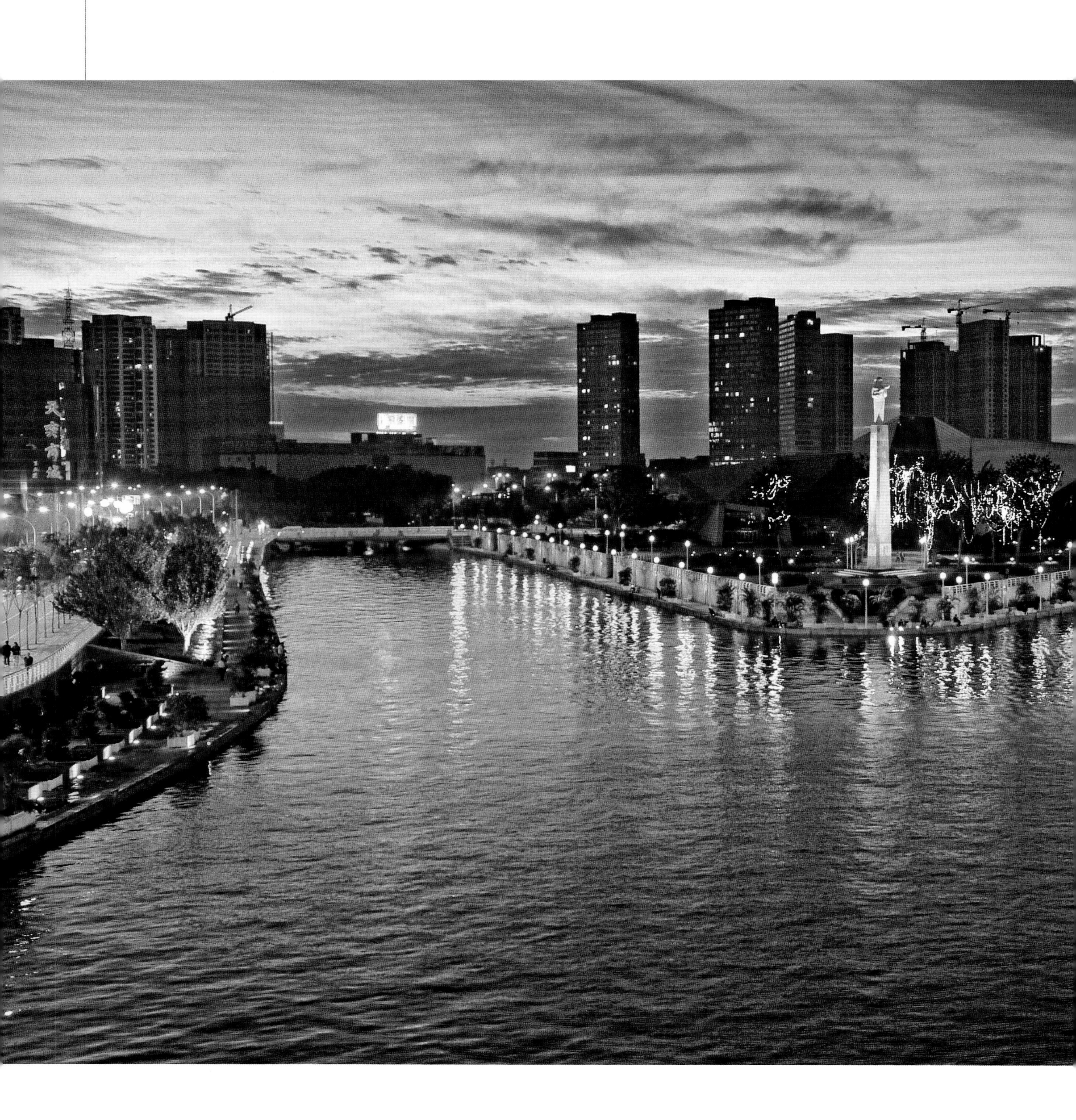

The Haihe River

The Haihe River is the general term for a collection of rivers running through North China and into Bohai Sea, including the Beiyun River, Yongding River, Daqing River, Ziya River, Nanyun River, Luanhe River and Tuhaimajia River. The catchment area is over 310,000 km^2, and the average annual runoff is 37.2 billion m^3.

珠　江

系西江、北江、东江和珠江三角洲诸河的总称。干流西江发源于云南马雄山，全长2 075公里。在我国境内的流域面积44万多平方公里，多年平均年径流量3 360亿立方米。

The Pearl River

The Pearl River is the general term for a collection of rivers including Xijiang River, Beijiang River, Dongjiang River, and other rivers in the Pearl River Delta. With a total length of 2,075 km, its mainstream originates in Maxiong Mountain in Yunnan Province. Its catchment area in China is over 440,000 km^2, and its average annual runoff is 336 billion m^3.

松花江

发源于中朝交界的长白山天池，是黑龙江在我国境内的最大支流，全长1 927公里，跨辽宁、吉林、黑龙江和内蒙古等四省、自治区，流域面积约55万平方公里，多年平均年径流量759亿立方米。

The Songhua River

Originating in Tianchi, Changbaishan Mountain on the boundary of China and the Democratic People's Republic of Korea, the Songhua River is the largest tributary of Heilongjiang River within China. The total river length is 1,927 km, running through Liaoning, Jilin, Heilongjiang Provinces, and Inner Mongolia Autonomous Region, covering a catchment area of 550,000 km^2. Its average annual runoff is 75.9 billion m^3.

辽 河

发源于河北平泉县，是我国东北地区南部最大的河流，全长1 345公里，流域面积22.9万平方公里，平均年径流量126亿立方米。

The Liaohe River

Originating in Pingquan County, Hebei Province, the Liaohe River is the largest river in the southern part of Northeast China. With a total length of 1,345 km, it covers a catchment area of 229,000 km^2, and the average annual runoff is 12.6 billion m^3.

鄱阳湖 *Poyang Lake*

位于江西北部，是我国第一大淡水湖，正常水位湖面面积3 150平方公里，容积为276亿立方米。

Situated in the north of Jiangxi Province, Poyang Lake is the largest freshwater lake in our country. The lake area at normal water level is 3,150 km^2, and the water volume totals 27.6 billion m^3.

洞庭湖 Dongting Lake

位于湖南北部，长江荆江河段以南，正常水位湖面面积2 820平方公里，容积为188亿立方米。

Dongting Lake is situated in the north of Hunan Province and to the south of Jingjiang Reach of the Yangtze River. The lake area at normal water level is 2,820 km^2, and the total water volume is 18.8 billion m^3.

洪泽湖 Hongze Lake

位于江苏洪泽县西部，正常水位湖面面积2 069平方公里，容积为26.6亿立方米。

Hongze Lake is situated in the west of Hongze County, Jiangsu Province. The lake area at normal water level is 2,069 km^2, and the total water volume is 2.66 billion m^3.

太湖　Taihu Lake

位于江苏和浙江两省交界处，正常水位湖面面积2 250平方公里，容积为51.5亿立方米。

Taihu Lake is situated at the boundary of Jiangsu and Zhejiang Provinces. The lake area at normal water level is 2,250 km^2, and the total water volume is 5.15 billion m^3.

巢湖 Chaohu Lake

位于安徽境内，正常水位湖面面积约750平方公里，容积为15.96亿立方米。

Chaohu Lake is situated in Anhui Province. The lake area at normal water level is 750 km^2, and the total water volume is 1.596 billion m^3.

青海湖　Qinghai Lake

位于青海东北部，正常水位湖面面积约4 583平方公里，容积为739亿立方米，是我国最大的内陆湖泊，也是我国最大的咸水湖。

Situated in the northeast of Qinghai Province, Qinghai Lake is the largest inland lake and the biggest saltwater lake in China. The lake area at normal water level is 4,583 km^2, and the total water volume is 73.9 billion m^3.

悠久的治水历史

中华民族的发展史，从某种意义上讲就是一部治水史。我们祖先留下了都江堰、灵渠、京杭运河等许多著名古代水利工程，至今仍发挥着重要作用，成为治水史上的光辉典范。

A Long History of Water Harnessing

The history of the development of the Chinese nation is to some extent a history of water harnessing. Our ancestors have bequeathed us many prestigious water conservancy projects including Dujiang Weir, Lingqu Canal and Beijing-Hangzhou Canal, which are playing important roles untill today, and have become great monuments in the history of water harnessing.

都江堰 *Dujiang Weir*

位于四川都江堰市，建于公元前256年左右，是世界上年代最为久远、仍在发挥显著效益、以无坝引水为特征的水利工程。

Located in Dujiangyan, Sichuan Province and built around 256 B.C., Dujiang Weir is the oldest consecutively operated water conservancy project in the world. It is still playing a significant role, featuring the technique of water diversion without dams.

灵渠 *Lingqu Canal*

位于广西桂林市境内，于公元前214年凿成通航，是世界上最古老的人工运河之一。

Located in Guilin, Guangxi Zhuang Autonomous Region, the Lingqu Canal was built and navigated in 214 B.C., and is one of the oldest artificial canal in the world.

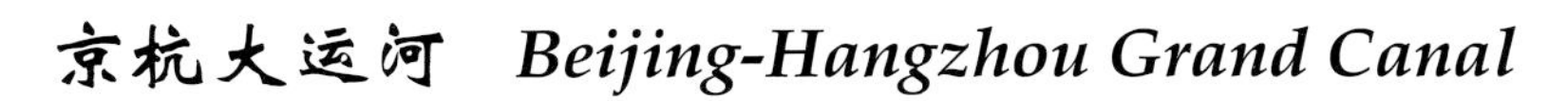

京杭大运河 *Beijing-Hangzhou Grand Canal*

最早开凿于公元前486年，1293年全线通航，全长1 794公里，是世界上里程最长、工程最大、最古老的运河。

First excavated in 486 B.C., Beijing-Hangzhou Grand Canal was completed and navigated in 1293. With a total length of 1,794 km, it is the longest, biggest and oldest artificial canal in the world.

水利建设成就辉煌

新中国成立60多年来，我国已建成江河堤防近30万公里，建成各类水库8.8万座，全国农田有效灌溉面积从新中国成立时的2.4亿亩增加到9.25亿亩，形成了较完善的水旱灾害防治与水资源供给工程保障体系，取得了举世瞩目的伟大成就。

Splendid Achievements of China's Water Conservancy Development

Since the founding of New China in 1949, the country has built nearly 300,000 km river dykes and 88,000 reservoirs of various types. The area of effective farmland irrigation has increased from 16 million hectares to the current figure of 61.7 million hectares. Remarkable achievements have been accomplished by building a relatively complete engineering system for flood-drought reduction and prevention, and water supply.

长江三峡水利枢纽

当今世界上最大的水利枢纽工程，总库容393亿立方米，防洪库容 221.5亿立方米，电站总装机容量2 250万千瓦，2012年发电量约为980亿千瓦时。通过三峡水库和长江堤防联合运用，可使荆江河段防洪标准达到100年一遇。

Three Gorges Water Control Project on the Yangtze River

Three Gorges Dam is the largest Water Control Project in the world, with a total capacity of 39.3 billion m^3, a flood control capacity of 22.15 billion m^3, a total installed capacity of 22.5 million kW. It generates an electricity output of about 98 billion kWh in 2012. The joint operation of the Three Gorges Dam and the Yangtze River embankment can raise the flood-control standards of the Jingjiang Reach to 100 years return period.

黄河小浪底水利枢纽

治理开发黄河的关键性工程，具有防洪、防凌、减淤、供水、灌溉、发电等综合效益。水库总库容126.5亿立方米，电站总装机容量180万千瓦。通过与三门峡、陆浑、故县水库联合运用，可使黄河下游防洪标准由60年一遇提高到1000年一遇。

Xiaolangdi Water Control Project on the Yellow River

Xiaolangdi Water Control Project is a key project in terms of the harnessing and development of the Yellow River with multiple functions in flood control, ice prevention, sediment reduction, water supply, irrigation and power generation. The total reservoir capacity is 12.65 billion m^3, and the total installed capacity is 1.8 million kW. Its joint operation with Sanmenxia, Luhun and Guxian reservoirs can raise the flood-control standards of the lower reaches of the Yellow River from 60 years return period to 1000 years return period.

淮河临淮岗水利枢纽

淮河干流迄今为止最大的水利枢纽工程，主体工程由主坝、南北副坝、引河、船闸、进泄洪闸等建筑物组成。该工程的建成结束了淮河中游无防洪控制性工程的历史，使淮河干流的防洪标准由过去的不足50年一遇提高到100年一遇。

Linhuaigang Water Control Project on the Huaihe River

Linhuaigang Water Control Project is by far the largest water conservancy project on the mainstream of the Huaihe River. The main part of the project consists of the main dam, auxiliary dam on the south and north, by-pass channel, ship lock, and incoming and discharge sluice. The completion of the Linhuaigang project was a milestone that ended the history of no flood-control engineering project in the middle reaches of the Huaihe River. The flood control standard of the Huaihe River was raised from less than 50 years return period to 100 years return period.

第二篇

Chapter Two

盛世兴水新篇

A New Chapter for Water Development in the New Century

2011年是我国水利发展史上具有里程碑意义的一年。中央出台一号文件，召开最高规格的中央水利工作会议，对加快水利改革发展作出全面部署，开启了中国特色水利现代化的新征程，掀开了中华民族治水兴邦的新篇章。2011年以来，水利建设呈现出前所未有的大好局面，水利发展改革成果普惠广大人民群众。

2011 was a landmark year in China's water development and management history. The No.1 Central Document and the highest level Central Working Conference on Water Resources made overall arrangements for the acceleration of water development and reform, initiating a new water modernization drive with Chinese characteristics, and opening a new chapter in water management and the development of the Chinese nation. Since 2011, water construction has experienced an unprecedentedly gratifying situation and the achievements of water development and reform have benefited a large number of people.

中共中央文件

中发〔2011〕1号

★

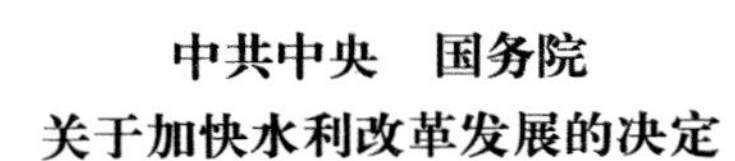

中共中央 国务院
关于加快水利改革发展的决定
（2010年12月31日）

水是生命之源、生产之要、生态之基。兴水利、除水害，事关人类生存、经济发展、社会进步，历来是治国安邦的大事。促进经济长期平稳较快发展和社会和谐稳定，夺取全面建设小康社会新胜利，必须下决心加快水利发展，切实增强水利支撑保障能力，实现水资源可持续利用。近年来我国频繁发生的严重水旱灾害，造成重大生命财产损失，暴露出农田水利等基础设施十分薄弱，必须大力加强水利建设。现就加快水利改革发展，作出如下决定。

— 1 —

2011年中央一号文件聚焦水利

2011年，中央一号文件首次聚焦水利。这是新中国成立以来，中央第一次出台综合性水利政策文件，第一次将水利提升到关系经济安全、生态安全、国家安全的战略高度，第一次鲜明提出水利具有很强的公益性、基础性和战略性特征。

No. 1 Central Document in 2011 Focusing on Water Conservancy

The No. 1 Central Document in 2011 was the first document of its kind focusing on water management. Since the founding of the People's Republic of China in 1949, it is the first time that the Central Committee of Communist Party of China issued a comprehensive water policy document. Furthermore, it was the first time that water was raised to a strategically high level relevant to economic, ecological and national security. Moreover, it was the first time that water was explicitly referred to in terms of public welfare, basic needs and strategic importance.

安徽黄山日出云海

Sun ascends from the sea of clouds at Huangshan Mountain, Anhui Province

中央召开高规格水利工作会议

2011年7月8－9日召开的中央水利工作会议，是中国共产党成立和新中国建立以来，首次以中央名义召开的水利工作会议，也是党的历史上最高规格的治水会议。会议明确了新形势下水利的重要地位，提出加快水利改革发展是关系中华民族生存和发展的长远大计。

The Highest Level Central Working Conference on Water Resources

The Central Working Conference on Water Resources, held on July 8-9th, 2011, was the first working Conference on water resources convened by the Central Committee of Communist Party of China since the founding of the People's Republic of China. The meeting underlined the great importance of water resource issues under the new situation and stressed that water development and reform is of long-term strategic significance to the development of the Chinese nation.

水利改革发展目标（2020年）

Water Reform and Development Goals (2020)

基本建成防洪抗旱减灾体系

To build a flood defense, drought relief and disaster mitigation system

重点城市和防洪保护区防洪能力明显提高，抗旱能力显著增强，“十二五”期间基本完成重点中小河流（包括大江大河支流、独流入海河流和内陆河流）重要河段治理、全面完成小型水库除险加固和山洪灾害易发区预警预报系统建设。

Flood defense capacities of key cities and flood defense zones will be substantially enhanced. Drought relief capacity will be greatly improved. During the 12th Five-Year Plan period (2011-2015), the treatment of key small and medium-sized rivers (including tributaries of large rivers, rivers running to the sea and inland rivers) will be completed, reinforcement of small-sized reservoirs will be carried out, and disaster forecasting and an early-warning system for flash floods in mountainous areas will be constructed.

基本建成水资源合理配置和高效利用体系

To establish a rational water resources allocation and efficient utilization system

全国年用水总量力争控制在6 700亿立方米以内，城乡供水保证率显著提高，城乡居民饮水安全得到全面保障，万元国内生产总值和万元工业增加值用水量明显降低，农田灌溉水有效利用系数提高到0.55以上，“十二五”期间新增农田有效灌溉面积4 000万亩。

Total annual national water consumption will not be allowed to exceed 670 billion m^3. The rural and urban water supply assurance rate will be substantially raised. Water consumed per 10,000 yuan of GDP and per 10,000 yuan of industrial added value will be remarkably reduced. The coefficient of irrigation water efficiency will be raised to above 0.55. During the 12th Five-Year Plan period, effective irrigation farmland will be increased by 2.66 million hectares.

基本建成水资源保护和河湖健康保障体系

To build a water resources protection, and river and lake health safeguards system

主要江河湖泊水功能区水质明显改善，城镇供水水源地水质全面达标，重点区域水土流失得到有效治理，地下水超采基本遏制。

The water quality of key water function zones will be greatly improved. Water quality of water sources for urban water supply will comprehensively meet its respective targets. Soil erosion in key areas will be effectively treated. Over-exploitation of groundwater will be curbed.

基本建成有利于水利科学发展的制度体系

To build an institutional system conducive to scientific development of water sector

最严格的水资源管理制度基本建立，水利投入稳定增长机制进一步完善，有利于水资源节约和合理配置的水价形成机制基本建立，水利工程良性运行机制基本形成。

The strictest water resources management system will be primarily established. Further measures will be taken to improve the stability of water investment system. A water pricing system conducive to saving and rational allocation of water will be primarily built. A sound water project operation system will be established.

万里长城
The Great Wall

新时期治水思路的丰富和完善

积极践行并不断丰富完善可持续发展治水思路，坚持以人为本，坚持人与自然和谐，坚持水资源可持续利用，坚持统筹兼顾，坚持改革创新，坚持现代化方向，积极推进传统水利向现代水利、可持续发展水利转变。

Enrich and Improve Water Management Theory

The government and its departments actively practice and continuously enrich and improve the sustainable water management concept, and insist on prioritizing people's welfare, harmony between human and nature, sustainable utilization of water resources, coordination, reform and innovation, and modernization, in order to facilitate the transition from traditional water management to modern and sustainable water management.

青海黄河九曲

The meandering Yellow River

从控制洪水向洪水管理转变

Transition from flood control to flood management

坚持科学防控、依法防控、综合防控，给洪水以出路，合理利用雨洪资源。

Insisting on comprehensive flood defense according to scientific and legal principles, to give room to flood water and rationally utilize flood water resources.

从供水管理向需水管理转变

Transition from water-supply management to water-demand management

坚持节水优先、以水定需、量水而行，全面建设节水型社会。

Insisting on prioritizing water saving, and managing water demand and development according to availability, so as to build a water-saving society in an all-round way.

从水土流失重点治理向预防保护、综合治理、生态修复相结合转变

Transition from the treatment of key water and erosion areas to prevention, protection, integrated treatment and ecological rehabilitation

坚持预防为主、保护优先，优化配置工程、生物和耕作措施，注重发挥大自然的自我修复能力。

Emphasizing prevention, prioritizing protection, rationalizing the application of engineering, biological and cultivation measures, and focusing on nature's ability to restore itself.

从水资源开发利用为主向开发保护并重转变

Transition from emphasizing water development and utilization to giving equal attention to development and protection

更加注重水利建设中的生态保护和移民安置，实现经济效益、社会效益和生态效益的多赢。

Paying more attention to ecological protection and the relocation of affected communities during project construction, in order to achieve a win-win situation with economic, social and ecological benefits.

民生水利理念与实践

近年来，水利部门把科学发展观的根本要求与民生水利的具体实践结合起来，以解决人民群众最关心、最直接、最现实的水利问题为重点，着力解决好直接关系人民群众生命安全、生活保障、生存发展、人居环境、合法权益等方面的民生水利问题，使人人共享水利发展与改革成果。

Concept and Practice of Water Management for People's Livelihood

In recent years, water management authorities have practiced managing water for people's livelihood according to the basic requirements of the Scientific Outlook on Development. While addressing the water issues that are of greatest concern to the general public, water management authorities intend to resolve the water problems that are most relevant to people's safety, lives, development, living environment and legal rights, to enable everyone share the benefits of water development and reform.

在防灾减灾中突出民生

Prioritizing the protection of people's livelihood during disaster prevention and mitigation

始终把保障人民群众的生命安全和饮水安全放在防汛抗旱工作的首位。

Insisting on giving first priority to people's safety and drinking water safety during disaster prevention and mitigation.

在水利建设中突出民生

Prioritizing the improvement of people's livelihoods during water conservancy construction

把人民群众直接受益的基础设施作为水利建设的优先领域。

Infrastructures that bring direct benefits to the general public given high priority in water conservancy construction.

在水利管理中突出民生

Prioritizing the improvement of people's livelihoods during water management

把维护群众的基本需求与合法权益放在水利管理的突出位置。

Satisfying people's basic needs and protecting people's legal rights given great attention during water resources management.

在水利改革中突出民生

Prioritizing the improvement of people's livelihood during water conservancy reform

在推进水利改革过程中切实保障群众的切身利益。

Effectively safeguarding people's interests during water sector reform.

云南普者黑

Puzhehei, Yunan Province

水利改革发展进入新阶段

2011年以来，神州大地治水兴水热潮涌动。从中央到地方，都将水利作为优先保障领域和重点支持对象，形成了治水兴水的强大合力。水利投资大幅增长，2011年水利建设投资3 452亿元，其中中央水利投资1 141亿元，首次突破1 000亿元；2012年，中央水利投资已突破1 500亿元。一大批民生水利工程开工建设，水利体制机制改革实现突破，最严格水资源管理制度扎实推进……

Water Reform and Development has Entered a New Phase

Since 2011, China has witnessed active water management and development. From the central to the local governments, water sector has been the priority area for government input, which forms strong synergies. Investment in the water sector has climbed substantially. Water construction investment in 2011 reached 345.2 billion yuan, out of which 114.1 billion yuan was from the central government, surpassing 100 billion yuan for the first time. In 2012, investment from central government rose to 150 billion yuan. A large number of water projects for people's livelihood are under construction. Institutional water reform has achieved breakthroughs. The implementation of the strictest water resource management system is under way...

西藏旁多水利枢纽夜间施工

Construction of Pangduo Water Control Project, Tibet Autonomous Region

第三篇

Chapter Three

保障防洪安全

Ensuring Flood Prevention Security

近年来，水利部门转变防汛工作思路，洪涝灾害防御能力和防御水平不断提高。通过坚持不懈地开展大规模防洪工程建设，大江大河主要河段可防御新中国成立以来最大洪水，以堤防、水库、蓄滞洪区等工程措施和防汛预警预报系统等非工程措施组成的综合防洪减灾体系初步形成，确保了大江大河、大中城市和重要基础设施的安全，确保了人民群众生命安全。

In recent years, the Ministry of Water Resources as well as departments in the water sector have updated their thinking on flood prevention and control, and continuously strengthened capabilities and standards in defending flood and water Logging disasters. Through unremittingly constructing flood prevention projects on a large scale, we are now able to guard ourselves against the worst floods since the founding of New China in the main sections of major rivers. Up to now, we have basically completed the establishment of a comprehensive flood prevention and disaster mitigation system consisting of both engineering measures (including dykes, reservoirs and flood detention areas) and non-engineering measures (including flood warning and forecasting systems), thus ensuring the safety of major rivers, large and medium-sized cities, essential infrastructure and people's life and property.

确立科学防洪理念

树立以人为本、生命至上、人水和谐，给洪水以出路的科学防洪理念，防汛思路由控制洪水、与洪水抗争向洪水管理、主动防御和疏导转变。注重实施洪水风险管理，科学调控洪水、依法调度洪水，将防治水患灾害与规范人类活动相结合。推行洪水资源化，落实以防为主、防抗结合的工作原则，建设统一指挥、协调有序的防汛组织体系，形成防灾抗灾的强大合力。

Establishing Scientific Thinking on Flood Control and Prevention

We have established scientific thinking on flood prevention and control which is people-oriented, and attaches the greatest importance to human life, harmony between human and water and gives room to floodwaters. We have changed our mindset from containing and combating floods to managing, preventing in advance and channeling floods. We emphasize implementing flood risk management, regulating and controlling floods scientifically, and schedule floods in line with the law, combining flood disaster prevention and the regulation of human activities. We have been promoting the practice of turning floods into a kind of resource, put in place the work principle of prevention first, combining the prevention and combating of floods. We strive to establish a coordinated and orderly flood prevention organizational system under a unified command, to effectively prepare for and combat disasters.

2011年台风“梅花”登陆山东荣成

The typhoon “plum” Landing in Rongcheng, Shandong Province, 2011

湖北丹江口水库大坝泄洪

Flood discharge of the Danjiangkou Dam, Hubei Province

大江大河大湖治理

1998年以来，新增堤防长度41 000余公里，长江下游河势控制、黄河标准化堤防建设稳步推进，治淮19项骨干工程全面竣工，长江三峡、嫩江尼尔基、广西百色、湖南皂市、黄河西霞院等一批重点水利枢纽建成投入运行，四川亭子口、江西峡江、广东乐昌峡、内蒙古海渤湾等水利枢纽工程开工建设。洞庭湖、鄱阳湖综合治理顺利实施，太湖流域水环境综合治理水利项目全面启动。

Harnessing of Major Rivers and Lakes

Since 1998, the length of newly-constructed dykes has amounted to over 41,000 km, while work in terms of river regimes control in the lower reaches of the Yangtze River and the construction of standard dykes along the Yellow River have been steadily pushed forward. A total of 19 key Huaihe River management projects have been completed; we have completed construction and put into operation a series of key water control projects such as Three Gorges Project in the Yangtze River, Neilki Reservoir in Nenjiang River, Baise Water Control Project of Guangxi Zhuang Autonomous Region, Zaoshi Reservoir of Hunan Province, Xixiayuan Water Control Project in the Yellow River; and the construction of several other key water control projects have been started already, including Tingzikou Water Control Project in Sichuan Province, Xiajiang Water Control Project in Jiangxi Province, Lechangxia Water Control Project in Guangdong Province, and Haibowan Water Control Project in Inner Mongolia Autonomous Region. The comprehensive management of Dongting Lake and Poyang Lake is being carried out in a smooth way, and the program for the comprehensive management of the aquatic environment of Taihu lake basin has been launched.

黄河标准化堤防
The Standard Dyke of the Yellow River

湖北荆江大堤
Jingjiang Dyke, Hubei Province

广西百色水利枢纽

Baise Water Control Project, Guangxi Zhuang Autonomous Region

江苏淮河入海水道

Sea-entering channel of the Huaihe River, Jiangsu Province

内蒙古尼尔基水库

Neilki Reservoir, Inner Mongolia Autonomous Region

防洪薄弱环节建设

2007年以来，累计完成86座大型、1 096座中型、16 251座小型病险水库的除险加固任务，大大消除了水库安全隐患，恢复防洪库容119.8亿立方米，恢复兴利库容155.3亿立方米，有效保障了水库下游2.8亿人、上千座县级以上城市、3.1亿亩农田以及大量重要基础设施的安全，进一步增强了水资源调控和抗御干旱灾害能力。

Reinforce Weak Links in Flood Prevention

Since 2007, we have reinforced 86 large, 1,096 medium and 16,251 small-sized hazardous reservoirs, thus greatly eliminating potential dangers of reservoirs. We have restored 11.98 billion m^3 of flood control capacity and 15.53 billion m^3 of utilizable capacity. As a result, the safety of 280 million people downstream of reservoirs, over one thousand cities above the county level, 20.7 million hectares of farmland and a large number of important infrastructure has been guaranteed, thus further enhancing the capability of water resources regulation and control as well as fighting and preventing drought disasters.

除险加固后的湖南官亭水库

Reinforced Guanting Reservoir, Hunan Province

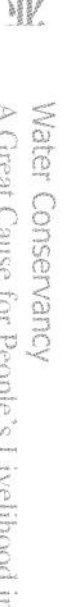

除险加固后的贵州格八水库

Reinforced Geba Reservoir, Guizhou Province

除险加固后的安徽佛子岭水库

Reinforced Foziling Reservoir, Anhui Province

从2009年起，启动了全国重点中小河流治理工程，以防洪保安为重点，着力提高重要河段防洪标准。已完成2 000多条重点中小河流重要河段治理任务，在防汛抗洪中发挥了重要作用。

山洪地质灾害防治非工程措施建设加快推进，通过制定防治规划、划定风险区域、健全应急预案、加强监测预警，建立预案到乡、预警到村、责任到人的山洪灾害防治机制，有效减少了人员伤亡和灾害损失。

Since 2009, we have launched the program to harness key small and medium-sized rivers nationwide, focusing on preventing floods and ensuring safety, with a view of improving flood prevention in major river sections. We have finished the training of key river sections of over 2,000 small and medium-sized rivers, which have played an essential role in flood prevention and control.

We have accelerated our efforts in applying non-engineering measures to prevent mountain torrent and geological disasters; we have effectively reduced casualties and losses caused by disasters by formulating prevention planning, demarcating risk areas, perfecting emergency preplans, strengthening monitoring and early warning as well as through establishing the mountain flood prevention mechanism featuring preplans at the township level, early warning at the village level and responsibilities delegated to individuals.

甘肃会宁县拦沙库

Sediment Retention Reservoir in Huining County, Gansu Province

北京密云县潮白河

Chaobai River in Miyun County, Beijing

水文建设

2005－2010年，共建设改造各类水文测站11 778个、水文巡测基地142个、水情（分）中心102个、水质监测分中心97个，遥感、地理信息系统(GIS)、全球定位系统(GPS)等现代信息技术的应用逐步扩展，水文测报技术明显改进。水文在科学防汛抗旱减灾、有效应对突发公共事件、加强水资源管理和水生态保护、为经济社会建设提供全面服务等方面发挥了重要作用。

Hydrological System Development

From 2005 to 2010, China constructed 11,778 Hydrometric stations of various kinds, 142 hydrological survey bases, 102 water regime centers, and 97 centers of water quality monitoring. The country has expanded the application of modern information technology such as remote sensing, GIS and GPS, and made notable progress in hydrological measuring and forecasting technologies. Hydrology has played an indispensable role in many areas such as scientific flood control and drought mitigation, effectively responding to public emergencies, strengthening the management of water resources and the protection of aquatic ecosystems and providing comprehensive services for socio-economic development.

浙江珊溪水库水位台

Water Level Station of Shanxi Reservoir, Zhejiang Province

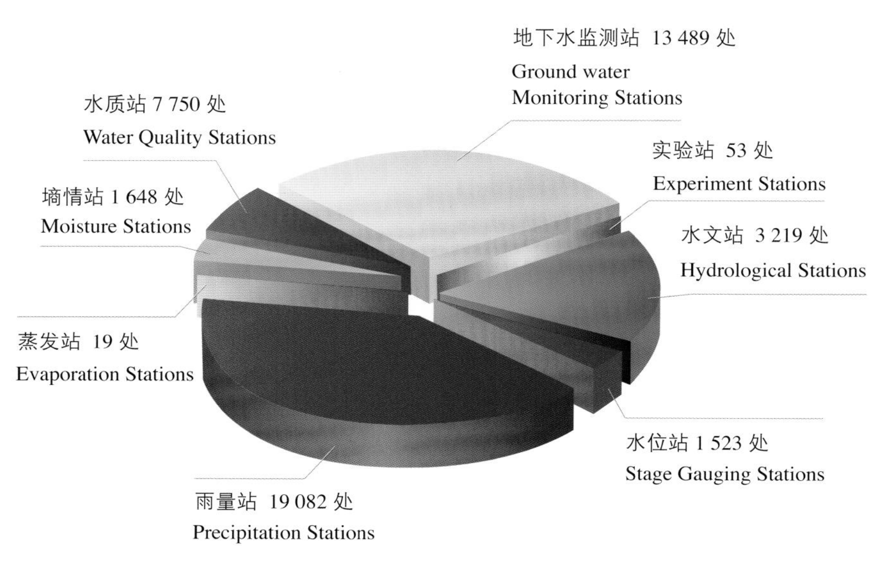

2011年各类水文测站构成图

Pie chart of various kinds of Hydrometric Stations, 2011

黄河花园口抢测洪峰

Emergency measurement of peak flow in Huayuankon section of the Yellow River

浙江上虞水文站

Shangyu Hydrological Station, Zhejiang Province

洪涝灾害防治成效显著

近年来，成功防御了频繁发生的洪水、台风和山洪灾害袭击，实现了大江大河安澜，保障了人民生命财产安全。2007年以来，全国防洪减灾的直接经济效益达8 289亿元，年均减淹耕地5 400万亩、减免粮食损失2 750万吨。

Remarkable Results Achieved in Prevention and Management of Flood and Water Logging Disasters

In recent years, we have taken successful defensive measures against frequent floods, typhoons and mountain torrents, restoring major rivers to their normal state and ensuring the safety of people and property. Since 2007, the direct economic benefits of national flood control and disaster mitigation work have amounted to 828.9 billion yuan. The average annual reduction of flooded farmland stands at 3.6 million hectares, and 27.5 million tons of grain has been saved every year.

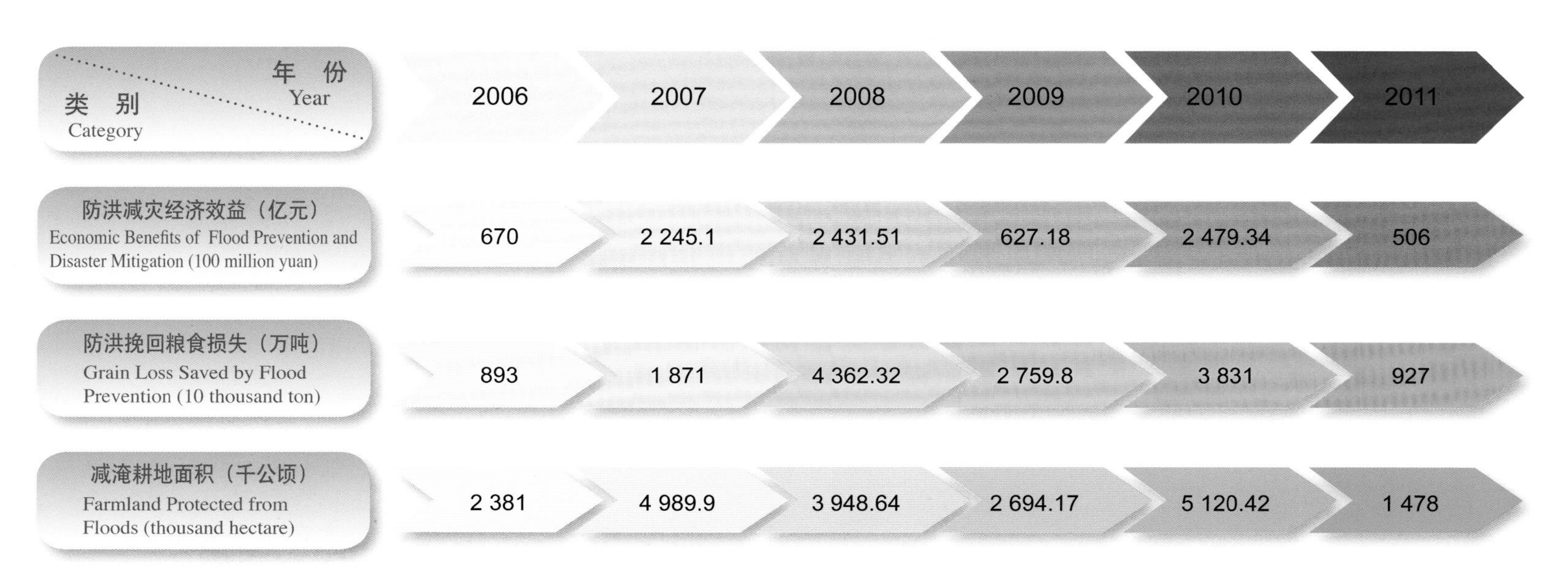

年份 Year / 类别 Category	2006	2007	2008	2009	2010	2011
防洪减灾经济效益（亿元） Economic Benefits of Flood Prevention and Disaster Mitigation (100 million yuan)	670	2 245.1	2 431.51	627.18	2 479.34	506
防洪挽回粮食损失（万吨） Grain Loss Saved by Flood Prevention (10 thousand ton)	893	1 871	4 362.32	2 759.8	3 831	927
减淹耕地面积（千公顷） Farmland Protected from Floods (thousand hectare)	2 381	4 989.9	3 948.64	2 694.17	5 120.42	1 478

2006—2011年我国防洪减灾效益统计

Statistics of national flood prevention and disaster mitigation benifits, 2006-2011

浙江缙云县练溪

Lianxi Stream of Jinyun County, Zhejiang Province

科学防控2007年淮河大水

2007年淮河发生了新中国成立以来仅次于1954年的流域性大洪水。国家防总、水利部科学调度上游水库拦蓄洪水，及时启动蒙洼等10处蓄滞洪区，充分利用下游入江入海水道排泄洪水，紧急转移80多万人。受灾地区无一人死亡、重要堤防无一处决口、水库无一座垮坝。

Scientific Prevention and Control of Extraordinary Flood in the Huaihe River in 2007

In 2007, a basin-wide flood occurred in the Huaihe River, second only to the one in 1954. The State Headquarters of Flood Control and Drought Relief and the Ministry of Water Resources regulated the upstream reservoirs to impound and retain the flood in a scientific manner, timely activating 10 flood detention areas, channeling the flood into rivers and the sea in the lower reaches of the Huaihe River, and evacuating 800,000 people. Therefore, we prevented the loss of a single resident in the flood-hit areas. None of the important dykes were breached, and no reservoir collapsed.

科学研判，果断决策

Scientific research and forecasting support decision-making

蒙洼蓄滞洪区王家坝开闸分洪

Flood diversion in Wangjiaba sluice of Mengwa flood retention area

淮河干堤

Main dyke of the Huaihe River

科学防御长江、黄河大水

2012年，长江上游发生1981年以来最大洪水，三峡水库迎来建库以来最大入库洪峰流量——71 200立方米每秒；黄河上游部分河段流量持续偏大，为近30年来之最。国家防总、水利部通过科学调度三峡工程拦蓄洪水，削减洪峰40%，有效减轻了长江中下游的防洪压力；通过加强黄河干流水库联合调度，确保了黄河安澜。

Scientific Defense against Major Floods in the Yangtze River and the Yellow River

In 2012, the largest flood since 1981 hit the upper reaches of the Yangtze River, sending the biggest peak flow into Three Gorges reservoir since its establishment—71,200 m^3/s; in some sections of the upper reaches of the Yellow River, the situation of high flow rate being bigger than normal persisted, which topped the past 30 years. State Flood Control and Drought Relief Headquarters and Ministry of Water Resources regulated Three Gorges Project to impound and detain flood, reducing peak flow by 40%, thus drastically reducing the pressure on the lower reaches of the Yangtze River. Through enhancing joint regulation of reservoirs in the main stream of the Yellow River, the Yellow River was restored to a tranquil situation.

新建成的宁夏黄河标准化堤防发挥防洪作用

The newly-constructed standard dyke of the Yellow River, Ningxia Hui Autonomous Region playing an impartant role in flood prevention

三峡水库泄洪

Flood discharge of Three Gorges Reservoir

有效防范汶川地震次生灾害

2008年5月12日，四川汶川发生8.0级特大地震。水利部门全力投入抗震救灾，与广大军民共同奋战，夺取了水利抗震救灾的重大胜利。灾区950多万人的饮水问题及时得到解决，2 400多座震损水库、800多座震损水电站无一垮坝，1 000多公里震损堤防无一决口，妥善化解了105处堰塞湖风险，实现了零伤亡。唐家山特大堰塞湖的安全除险，创造了世界上成功处置大型堰塞湖的奇迹。

Effective Prevention of Secondary Disasters in the Aftermath of Wenchuan Earthquake

On May 12th, 2008, a devastating earthquake measuring 8.0 on the Richter scale struck Wenchuan in Sichuan Province. Departments and agencies in the water sector made all-out efforts to combate the earthquake and disaster relief. They fought the hard battle together with the people and soldiers and scored a tremendous victory in combating the earthquake and disaster relief. The drinking water problem of more than 9.5 million people in the disaster-stricken area was resolved in a timely manner; neither of the over 2,400 earthquake-damaged reservoirs nor more than 800 earthquake-hit hydropower stations collapsed; not a single section of more than 1,000 km of earthquake-damaged dykes was breached. We successfully eliminated 105 potential dangers at dammed lakes and no casualties were witnessed. Through overcoming risks at the Tangjiashan dammed lake, we achieved a great victory in terms of successfully handling big dammed lakes.

水利部部长、水利部抗震救灾指挥部总指挥陈雷在唐家山堰塞体上主持召开会议，现场研究堰塞湖应急处置方案

H.E. Mr. Chen Lei, Minister of Water Resources and General Commander of the MWR Headquaters of Earthquake Rescue and Relief, is holding a meeting on the main barrier body of Tangjiashan, to formulate the emergeney plan of coping with the barrier lake

米-26运输机源源不断地将抢险设备吊运至抢险现场
An M-26 helicopter flew around the clock to lift rescue equipment to the site

紧急开挖导流明渠
Emergent excavation of a diversion channel

2008年6月10日，唐家山堰塞湖处置取得决定性胜利。图为导流明渠泄流状况

On June 10th, 2008, the decisive victory was scored in the danger-elimination of Tangjiashan Barrier Lake. The water dammed by the barrier is discharged through the diversion channel

水利部紧急向灾区调运抢险物资

The MWR is transporting emergency relief materials to the disaster-stricken areas

水利抢险队进行堤防现场抢险查勘

The water conservancy rescue team is investigating the damaged dykes

水利部应急水源水质监测组现场检测水质

The MWR emergency water source and water quality monitoring team is testing water quality on the site

妥善处置舟曲特大泥石流灾害

2010年8月8日，甘肃舟曲发生特大山洪泥石流灾害，大量冲积物淤堵白龙江形成堰塞湖，1/3以上的城区被淹，最深处达10米。水利部门第一时间派出工作组赶赴现场，全力以赴组织抢险。按照安全、科学、迅速的原则，采取挖、爆、冲等措施，与解放军、武警部队一道连续奋战24天，圆满完成舟曲堰塞湖应急处置及白龙江淤堵河段清淤疏通任务。

Proper Handling of the Extraordinary Debris Flow Disaster in Zhouqu County, Gansu Province

On August 8th, 2010, a tremendous mountain torrent and mud-flow disaster struck Zhouqu County in Gansu Province. A large dammed lake was formed due to the huge amount of alluvial deposits clogging the Bailong River. Over one-third of the city was flooded, with the maximum depth of 10 meters. Water conservancy departments quickly sent work teams to the site to help organize rescue work. Guided by the principle of carrying out rescue work in a safe, scientific and fast manner and implementing such measures as excavation, explosions and flushing with water, our water conservancy staff fought a hard battle together with soldiers of the People's Liberation Army and the Armed Police Force for 24 days. Finally, we achieved a complete success in emergency response work at the dammed lake in Zhouqu and the dredging of clogged sections of the Bailong River.

对白龙江泥石流淤塞体进行开挖

Excavation of the debris flow blocking the Bailong River

白龙江爆破疏通

Detonation of the debris body in the Bailong River

白龙江清淤

Dredging the Bailong River

夜战白龙江

Excavation operation deep into the night along the Bailong River

经过20多天的应急处置，县城受淹区域全面退水

After over 20 days of emergency operation, water receded in the flooded areas in the county

第四篇

Chapter Four

保障供水安全

Ensuring Water Supply Security

根据水资源承载能力和经济社会发展需求，2006年以来，兴建了一大批水资源开发利用工程，新增年供水能力285亿立方米，全国水利工程年供水能力达到7 000多亿立方米，城乡供水保障率显著提高。大力推进农村饮水安全工程建设，提前6年实现了联合国千年宣言提出的目标。世界上施工规模最大、供水规模最大、受益人口最多的调水工程——南水北调工程顺利实施，水资源调度工作不断加强，江河湖库水系连通工程加快建设。

In accordance with the carrying capacity of water resources, as well as the demand of the economy and society, a great number of water resource exploitation and utilization projects have been constructed since 2006, increasing the annual water supply by 28.5 billion m^3. The national water supply capacity stands at 700 billion m^3. Water supply capacity has been significantly improved in both urban and rural areas. With immense efforts to construct drinking water safety projects in rural areas, the goal of the United Nations Millennium Declaration has been achieved six years ahead of schedule. Construction of the South-to-North Water Diversion Project, world's largest water transfer project in terms of construction scale, water supply capacity and benefited population, has been progressing smoothly. Efforts have been made to strengthen water regulation and interconnection among rivers, lakes and reservoirs.

保障农村饮水安全

建设各类农村集中式供水工程，受益人口比例由2005年的40%提高到2011年的63%。截至2011年底，有2.76亿农村群众和2 600多万农村学校师生喝上了放心水。

Safeguarding Rural Drinking Water Safety

With the construction of various centralized water supply projects in rural areas, the proportion of the beneficiary population has risen from 40% in 2005 to 63% in 2011. As of the end of 2011, safe drinking water project has benefited 276 million rural residents, including 26 million rural teachers and students.

甘肃东乡族自治县少数民族群众喜迎“幸福水”

The ethnic minority people taste ‘the Water of Happiness’ in Dongxiang Autonomous County, Gansu Province

河北蓟县孩子们喝上了纯净的自来水

The children drink pure tap water in Ji County, Hebei Province

宁夏固原市某蓄水池

A pool for water supply in Guyuan, Ningxia Hui Autonomous Region

保障城镇供水

在城镇化快速发展的背景下，加大城镇水源工程挖潜改造，提升蓄供水能力。综合考虑城镇发展和应对极端天气、水源污染等突发事件的水源保障需求，加快城镇新水源和相关供水设施建设，因地制宜优化水源结构。加快城镇供水设施和管网改造，确保城镇供水安全。

Safeguarding Urban Water Supply

With the rapid extension of urbanization, water resource projects have been improved and reconstructed, to increase the capacity of water storage and supply. With overall consideration to safeguarding water supply against the effects of fast urbanization, extreme weather conditions and water pollution, new water resources in urban areas have been explored and supply facilities have been constructed to optimize water resources according to local conditions. Water supply facilities and the pipeline network have been upgraded to ensure urban water supply.

陕西水质化验中心

Water Quality Inspection Center, Shaanxi Province

广东潮州供水枢纽西溪拦河闸

Xixi Sluice of Water Supply Project in Chaozhou, Guangdong Province

城乡供水一体化——浙江诸暨市城南水厂

Integration of urban and rural water supply: Chengnan Water Plant in Zhuji, Zhejiang Province

水源工程建设

为解决资源性缺水和工程性缺水问题，开工建设了甘肃引洮、贵州黔中、广西壮族自治区乐滩、西藏自治区旁多、吉林哈达山等一批重点水源工程，加快推进西南中型水库建设，基本完成了陕甘宁盐环定扶贫扬黄续建任务。

甘肃贺家湾水库

Hejiawan Reservoir, Gansu Province

Construction of Water Source Projects

To solve the problem of water shortages due to insufficient water resources and underdeveloped infrastructure, some key water sources projects have been constructed in Yinzhao, Gansu Province, Qianzhong in Guizhou Province, Letan in Guangxi Zhuang Autonomous Region, Pangduo in Tibet Autonomous Region and Hadashan in Jilin Province. The construction of medium-sized reservoirs in Southwest China has been enhanced and Yanhuanding Pumping Project has been completed.

上海青草沙水库

Qingcaosha Reservoir, Shanghai

北京密云水库

Miyun Reservoir, Beijing

跨流域调水工程建设

实施跨流域调水工程，南水北调东、中线一期工程稳步推进，京石段建成发挥效益，对保障首都供水安全发挥了重要作用。

Construction of Inter-Basin Water Transfer Projects

The first phase of the east route and middle route construction of the South-to-North Water Diversion Project is well under way. The Beijing-Shijiazhuang section of the Diversion Project is playing an important role in securing water supply in Beijing.

南水北调中线工程——湖北丹江口水库

Middle route of South-to-North Water Diversion Project: Danjiangkou Reservoir, Hubei Province

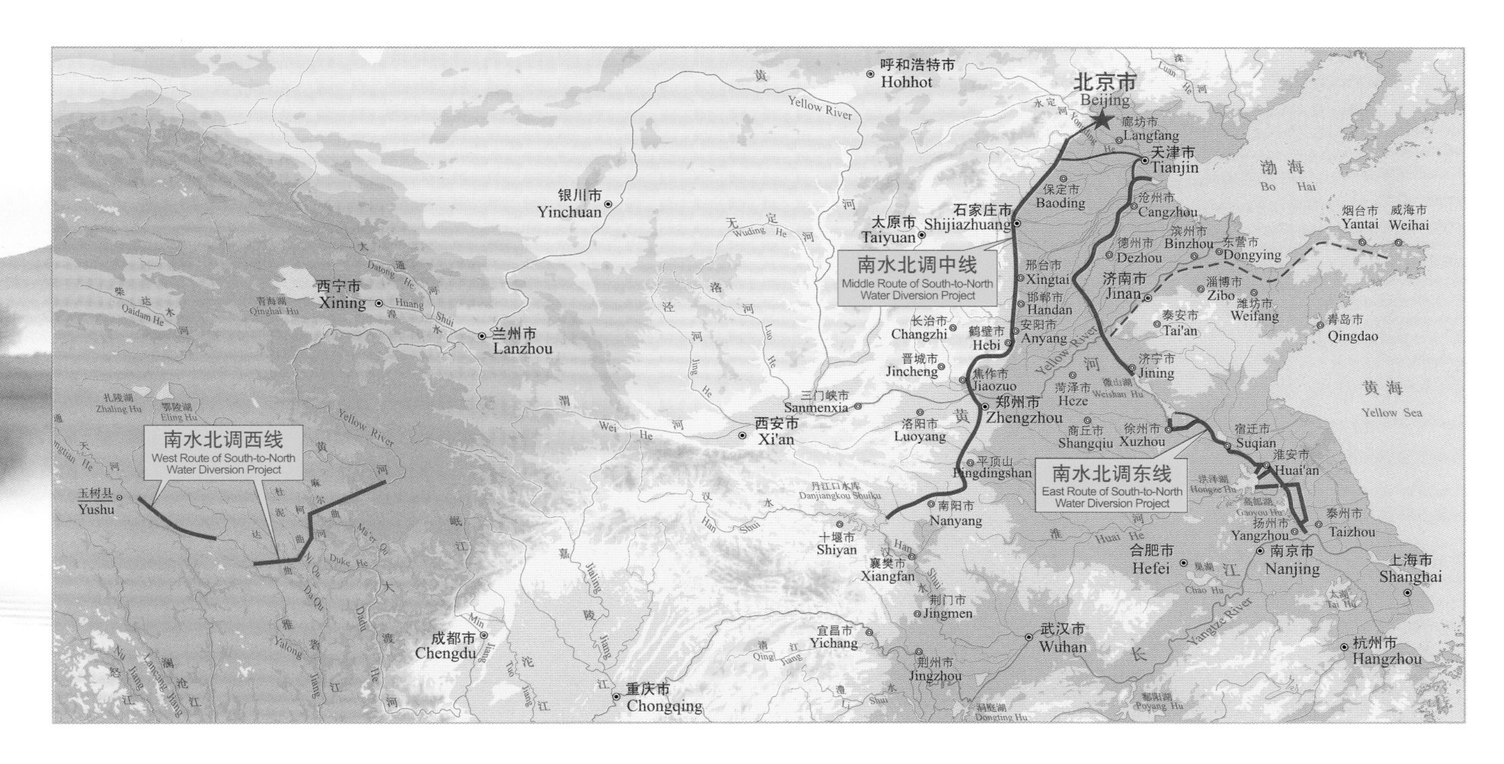

南水北调东、中、西线工程示意图

East route, middle route and west route of South-to-North Water Diversion Project

南水北调东线工程——泰州引江河枢纽

The Yinjiang Canal in Taizhou, Jiansu Province, part of the east route of the South-to-North Water Diversion Project

江河湖库水系连通工程建设

建设水库、闸坝、泵站、渠道等工程，综合采取调水引流、清淤疏浚、生态修复等措施，科学合理建设河湖水系连通工程，构建引排顺畅、蓄泄得当、丰枯调剂、多源互补、调控自如的江河湖库水网体系，进一步优化水资源配置格局。

Project Construction for the Interconnection of Rivers, Lakes and Reservoirs

Reservoirs, dykes, pumping stations and channels have been constructed. Water transfer, dredging and ecological rehabilitation have been implemented. By connecting rivers, lakes and reservoirs, a water network with unhindered channeling and drainage, proper storage and discharge, balanced water fluctuation, supplemented by various resources, and flexible regulation has been established to optimize the water resources distribution.

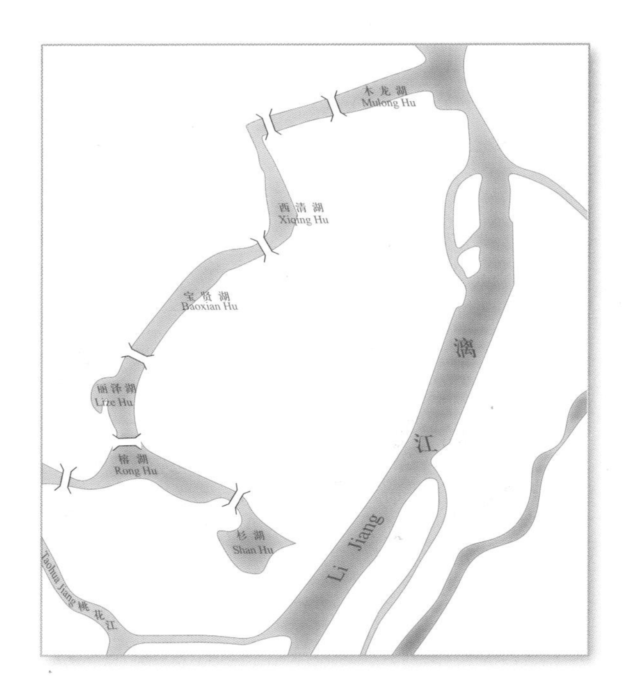

广西桂林两江四湖连通示意图

Interconnection project among two rivers and four lakes in Guilin, Guangxi Zhuang Autonomous Region

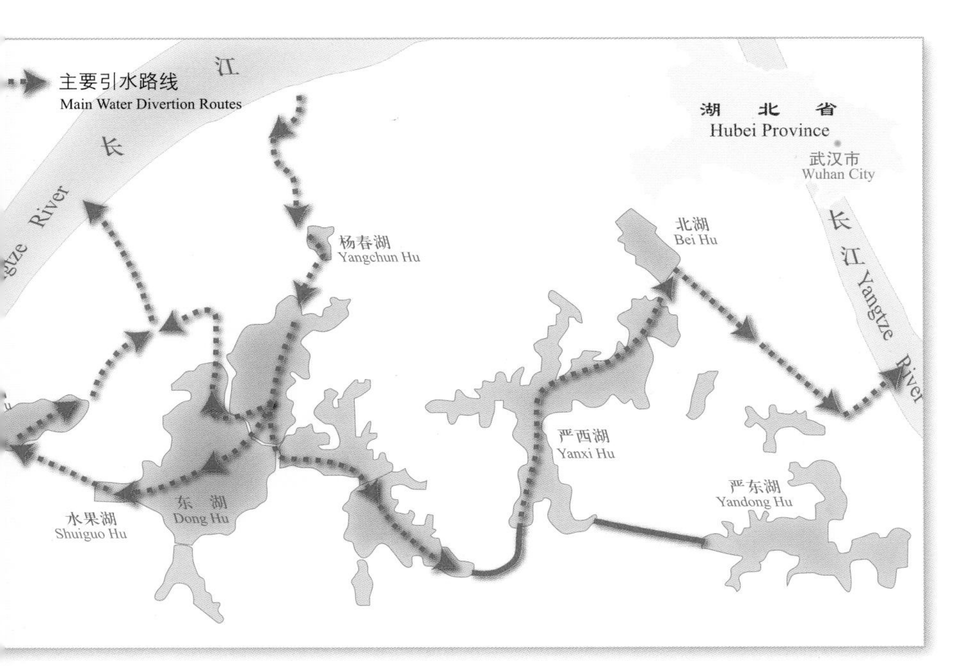

武汉东湖——六湖连通示意图

Interconnection project among six lakes: Donghu Lake, Wuhan

武汉东湖

Donghu Lake, Wuhan City

广西桂林打造河湖连通工程

Interconnection project among rivers and lakes in Guilin, Guangxi Zhuang Autonomous Region

有效应对西南地区持续严重干旱

2009—2011年，我国西南地区连续三年发生大旱，降雨量较常年减少四成以上。国家防总、水利部及时启动应急预案，科学调配抗旱水源，与广大军民奋力抗灾，通过新建应急水源、抢打机井、拉水送水等措施，解决了5 557万群众饮水困难，确保群众都能喝上水，最大程度地减轻了旱灾影响和损失。

Effectively Coping with Persistent and Severe Droughts in Southwest China

From 2009 to 2011, severe droughts took place in Southwest China for three consecutive years. Rainfall was reduced by 40% compared to regular levels. The State Flood Control and Drought Relief Headquarters and the Ministry of Water Resources launched a contingency plan to allocate water resources to combat the drought. With combined efforts from the People's Liberation Army and the people, through establishing contingency water resources, digging well and water delivery, 55.57 million people obtained access to drinking water, and losses from the drought were minimized.

贵州安顺苗族自治县通水缓解旱情

People get access to water in drought season in Anshun Miao Autonomous County, Guizhou Province

解放军为云南石屏县小学生分水

The PLA soldiers distribute water to pupils in Shipin County, Yunnan Province

广西西林县群众喜迎水井出水

Water comes out from a well in Xilin County, Guangxi Zhuang Automonous Region

实施应急调水

为解决北京、天津、澳门等地的用水困难，近年来组织实施了河北山西向北京输水、三峡水库向中下游补水、珠江枯水期水量调度、引黄济津济冀济淀等多次应急调水，确保了用水安全和社会稳定。

Implementing Emergency Water Transfer

To solve the problem of water shortage in Beijing, Tianjin and Macau, some emergency water transfer projects have been implemented, including water transfer from Hebei and Shanxi provinces to Beijing, water supplement from the Three Gorges Reservoir to the mid and lower reaches of the Yangtze River, water transfer to the Pearl River during the drought period, and water transfer from the Yellow River to Tianjin, Hebei Province and Baiyangdian Lake, to ensure water supply and social stability.

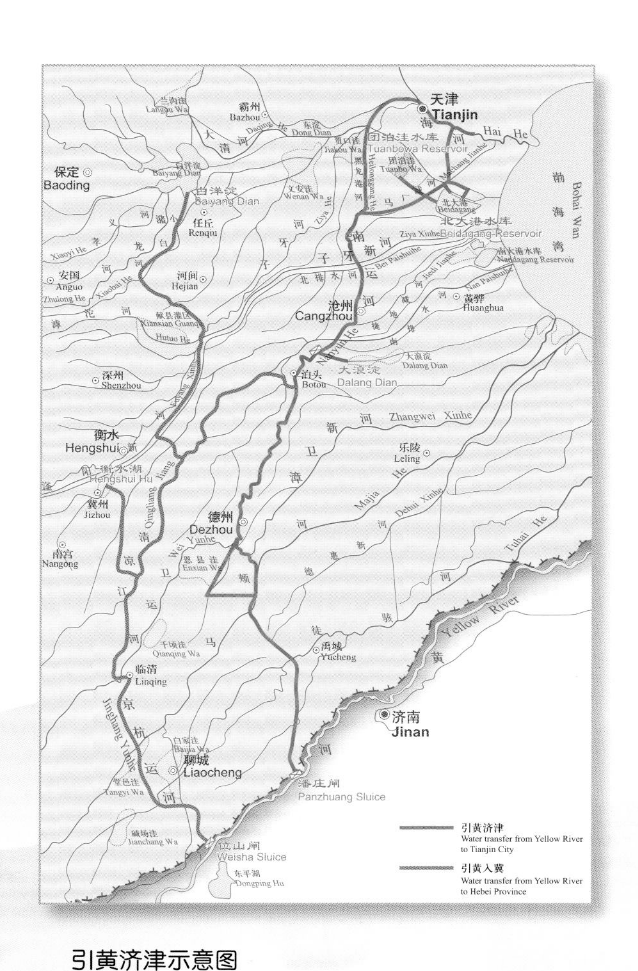

引黄济津示意图

Water transfer from the Yellow River to Tianjin

实施应急调水后的白洋淀

Baiyangdian Lake after emergency water transfer

珠江压咸补淡工程广州段

Guangzhou segment of Pearl River Water Transfer Project for repelling saltwater intrusion and supplementing freshwater

向北京输水的河北黄壁庄水库

Huangbizhuang Reservoir in Hebei Province, from which water is transferred to Beijing

第五篇

Chapter Five

保障粮食安全

Ensuring Food Security

水利兴，五谷丰。保障粮食安全，水利是基础和关键。经过多年大规模农田水利建设，建成各类农田水利工程2 000多万处，全国农田有效灌溉面积从新中国成立时的2.4亿亩增加到9.25亿亩，占世界总量的1/5，居世界首位。2012年我国粮食实现了自2004年以来的“九连增”。我国以占世界6%的淡水资源、9%的耕地，保障了约占全球21%人口的吃饭问题，为世界粮食安全作出了巨大贡献。

Water conservancy is the basis of and key to safeguarding food security. Over 20 million rural water projects have been constructed. Effective irrigation areas across the country have been increased from 16 million hectares in 1949 to 61.7 million hectares, accounting for one-fifth of the world's total and ranking first in the world. From 2004 to 2012, agricultural output increased for nine consecutive years. China feeds 21% of the world's population with only 6% of the world's freshwater and 9% of its arable land, thus making a great contribution to global food security.

大中型灌区续建配套与节水改造

2002年以来，对426处大型灌区和376处重点中型灌区骨干工程进行了续建配套与节水改造。新建了嫩江尼尔基、宁夏回族自治区沙坡头、海南大广坝、四川武都等一批灌区。近五年来，全国净增灌溉面积5 600多万亩，改善灌溉面积1.9亿亩，新增粮食综合生产能力约400亿斤，新增节水能力210亿立方米。

Water-Saving Renovation and Construction of Supporting Facilities in Large and Medium-Sized Irrigation Districts

Since 2002, reconstruction of water-saving facilities has taken place in 426 large irrigation districts and 376 key medium-sized irrigation districts. Some new irrigation districts were constructed such as Neilki of Nenjiang River, Shapotou in Ningxia Hui Autonomous Region, Daguang Dam in Hainan Province, and Wudu in Sichuan Province. In the past five years, the national irrigation area has increased by 3.73 million hectares, and the improved irrigation area reached 12.67 million hectares, while food productivity increased by 20 billion kg, and water saving capacity increased by 21 billion m^3.

四川都江堰灌区——我国特大型灌区之一，有效灌溉面积1 035万亩

Dujiang Weir Irrigation District in Sichuan Province, one of the mega irrigation districts in China, with effective Irrigation areas of 0.69 million hectares

内蒙古河套灌区——我国特大型灌区之一，有效灌溉面积860万亩

Hetao Irrigation District in Inner Mongolia Autonomous Region, one of the mega irrigation districts in China, has effective irrigation areas of 0.57 million hectares

安徽淠史杭灌区——我国特大型灌区之一，有效灌溉面积1 000万亩

Pishihang Irrigation District in Anhui Province, one of the mega irrigation districts in China, has effective Irrigation areas of 0.67 million hectares

西藏曲水农田

The farm in Qushui, Tibet Autonomous Region

湖南铁山灌区桃林渡槽

Taolin Aqueduct in Tieshan Irrigation District, Hunan Province

四川武都引水渠

Wudu Diversion Canal, Sichuan Province

宁夏固海引黄泵站

Guhai Pumping Station transferring water from the Yellow River, Ningxia Hui Autonomous Region

大型灌排泵站更新改造

从2009年开始，对全国大型灌排泵站实施更新改造，解决了泵站能耗高、灌排效益差等突出问题，有效改善了农业生产和群众生活条件。

Upgrading and Renovation of Large Pumping Stations for Irrigation and Drainage

Since 2009, large pumping stations for irrigation and drainage have been upgraded to solve the problems of high consumption of energy and low efficiency in irrigation and drainage, improving agricultural production and people's livelihood.

江苏江都排灌站

Jiangdu Pumping Station for irrigation and drainage, Jiangsu Province

病险水闸除险加固

2011年起，组织开展全国大中型病险水闸除险加固工作，确保长期老化失修的水闸能够正常发挥其防洪、排涝和兴利等功能，消除防洪安全隐患。

Hazards-removing and Reinforcement of Defective Water Sluices

Since 2011, reinforcement of defective large and medium-sized water gates has been implemented, to ensure that aging water sluices can properly function for flood protection, drainage and water utilization, and to eliminate the threat of flooding.

淮河蒙洼蓄洪区王家坝闸

Wangjiaba Sluice in Mengwa Flood Detention Area of the Huaihe River

高效节水灌溉

我国在连续30多年灌溉用水总量保持零增长的情况下，节水灌溉工程面积由2000年的2.46亿亩增加到2011年的4.38亿亩，农田灌溉水有效利用系数由0.3提高到0.51，亩均灌溉用水量由479立方米下降到367立方米，粮食总产量增加了5 000多亿斤。

High Efficiency Water-Saving Irrigation

While the total water consumption of irrigation has maintained zero growth for the past 30 consecutive years, the water-saving irrigation area has increased from 16.4 million hectares in 2000 to 29.2 million hectares in 2011; the effective utilization coefficient of farmland irrigation has been raised from 0.3 to 0.51; the average irrigation water consumption per mu (0.067 ha) has dropped from 479 m^3 to 367 m^3; and total grain production has increased by more than 250 billion kg.

宁夏大型喷灌设施

Large sprinkling irrigation facilities, Ningxia Hui Autonomous Region

小型农田水利重点县建设

2009年，启动实施小型农田水利重点县建设，到2011年底覆盖了1 250个重点县，项目区共发展低压管道输水、喷灌、微灌等高效节水灌溉面积1 700万亩，新增、恢复有效灌溉面积4 100万亩，改善灌溉面积5 400万亩，新增粮食生产能力308多亿斤。

Water Conservancy Projects for Small-Sized Farmland in Key Counties

In 2009, water conservancy projects for small-sized farmland were launched in key counties. By 2011, 1,250 counties had a water-saving irrigation area of 1.13 million hectares using low-pressure delivery pipelines, sprinkler irrigation and micro irrigation. The new and restored effective irrigation area has reached 2.73 million hectares, while 3.6 million hectares of irrigation area has been improved, and the grain production capacity has increased by 15.4 billion kg.

四川大邑县田间农渠建设

Irrigation ditches in the field of Dayi County, Sichuan Province

浙江江山市峡口水库灌区主干渠

Main canal in Xiakou Reservoir Irrigation District in Jiangshan, Zhejiang Province

东北四省区节水增粮行动

2012年，启动东北四省区节水增粮行动。计划用四年时间，投资380亿元，在东北四省区集中连片建设3 800万亩高效节水灌溉精品工程，新增粮食综合生产能力200亿斤，年均增收160多亿元，年均节水29亿立方米，农田灌溉水有效利用系数达到0.80以上。

Water-Saving and Grain Output-Boosting Programme in Four Northeast Provinces

In 2012, the Water-Saving and Grain Output-Boosting Programme was implemented in four provinces in Northeast China. In four years, with a total investment of 38 billion yuan, the flagship water saving irrigation project will be completed, covering 2.53 million hectares in Northeast China. The project is expected to generate 10 billion kg of increased grain production capacity, 16 billion yuan of annual increased income and 2.9 billion m^3 of annual water saving volume. The effective utilization coefficient of farmland irrigation is expected to be over 0.80.

东北粮食作物区

Crops area in Northeast China

第六篇

Chapter Six

保障生态安全

Ensuring Ecological Security

昔日荒山秃岭、田瘦人穷，今朝鸟语花香、水丰民富。我国是世界上水土流失最为严重的国家之一。近年来，水利部门将预防、保护、监督、治理和修复有机结合，优化配置工程、生物和耕作措施，构建科学完善的水土流失综合防治体系。通过治理，改善了生态环境，大幅度提高了土地生产力，改变了水土流失地区贫穷落后的面貌。同时，积极开展水生态系统保护与修复，大力推进农村水电开发利用，加强牧区水利建设，有效促进了生态环境的修复和改善。

In the past, people were trapped in poverty due to barren mountains, eroded hills and infertile lands. People now enjoy a prosperous life with abundant water resources. China's situation with regard to soil erosion and water loss is one of the most serious in the world. In recent years, water administration authorities integrated prevention, protection, supervision with regulation and restoration, and optimized structural, biological and farming measures to construct a scientific and comprehensive system for soil and water conservation. Regulatory efforts have improved the ecological environment, greatly increased land productivity, and ended the poverty and backwardness in areas suffering soil erosion and water loss. In the meantime, vigorous efforts are being made to protect and restore the water ecosystem, energetically promote the development and utilization of hydropower in rural areas, and intensify water conservancy construction in pasture areas, effectively facilitating the restoration and improvement of the ecological environment.

水土保持

通过小流域综合治理、坡耕地整治、淤地坝建设、生态自然修复以及开展重点区域治理等措施，截至2011年底，全国累计治理水土流失面积110万平方公里。在600多个县开展重点县治理，每年可减少土壤侵蚀量15亿吨以上，增加蓄水保水能力250多亿立方米，增产粮食360亿斤。

Soil and Water Conservation

By the end of 2011, soil erosion and water loss in an accumulated area of 1.1 million km^2 had been controlled by measures such as integrated small watershed management, slope farmland regulation, warping dam construction, natural ecological restoration, and regulation in key areas. Soil and water conservation efforts in more than 600 key counties can reduce soil erosion amount by more than 1.5 billion tons, increase water storage and conservation capacity by over 25 billion m^3, and raise grain output by 18 billion kg per annum.

福建龙岩长汀县桐坝后山水土保持

Soil and water conservation in Tongba Mountain, Changting County, Longyan, Fujian Province

北京门头沟区清洁小流域

Clean small watershed in Mentougou District, Beijing

内蒙古阿尔山封山禁牧

Mountain enclosure and grazing prohibition in A'ershan Mountain, Inner Mongolia Autonomous Region

宁夏彭阳县梯田

Terraced fields in Pengyang County, Ningxia Hui Autonomous Region

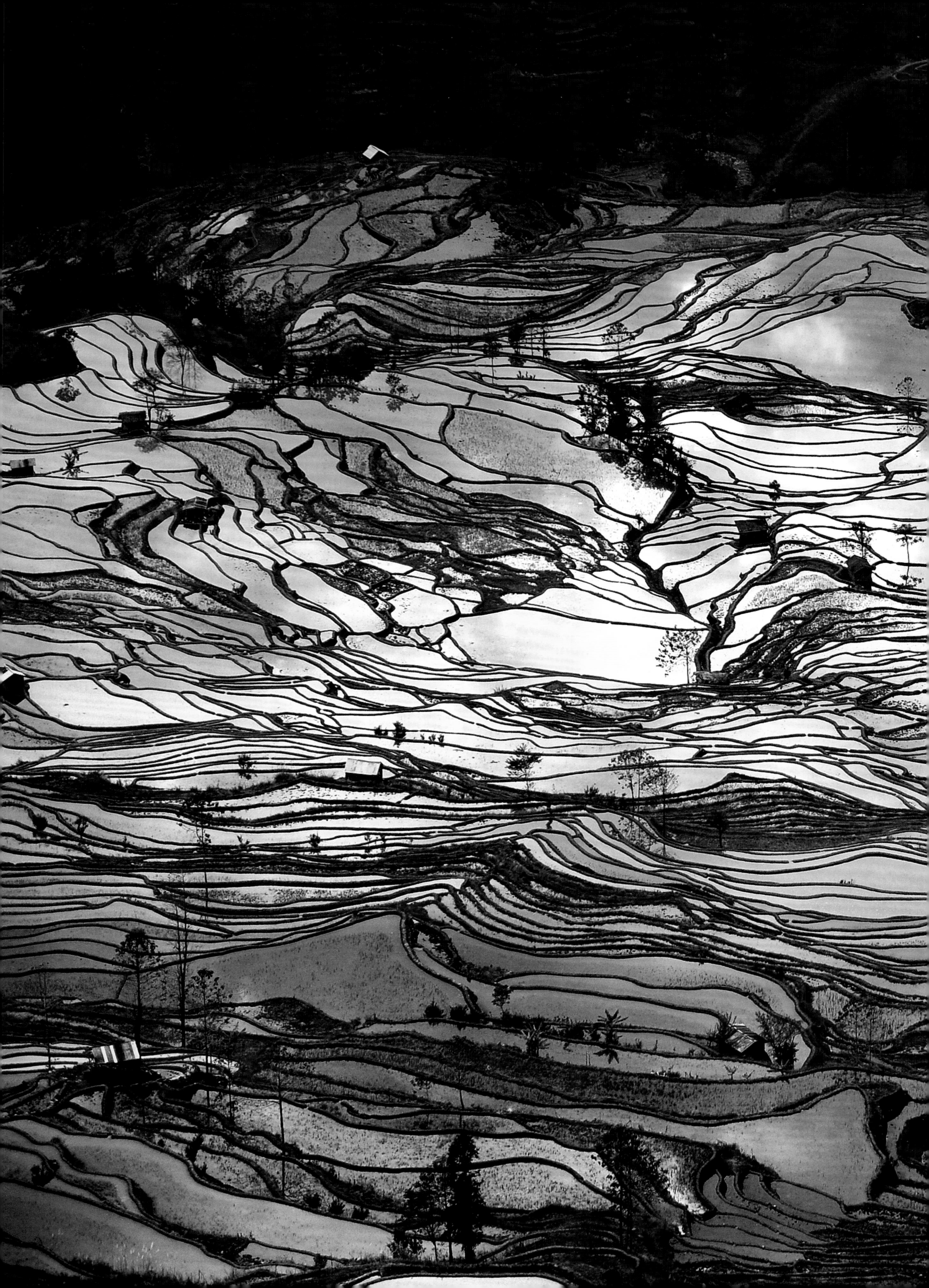

云南元阳梯田

Terraced fields in Yuanyang County, Yunnan Province

水生态系统保护与修复

近年来，开展了大量水生态系统保护与修复的实践和探索，先后对黄河、塔里木河、黑河、石羊河等生态脆弱流域实行水资源统一调度和综合治理，黄河连续13年不断流，塔里木河、黑河、石羊河下游生态得到很大改善，引黄济淀、扎龙湿地补水等应急生态调水取得了明显效果。

Protection and Restoration of Water Ecosystem

In recent years, tremendous efforts have been made to practice and explore the protection and restoration of water ecosystem by implementing unified allocation and integrated management of water resources in ecologically fragile river basins such as the Yellow, Tarim, Heihe, and Shiyang rivers. Sections of the Yellow River stopped drying up for 13 consecutive years, the ecological condition in Tarim River, Heihe River, and downstream Shiyang River greatly improved, emergency response ecological water diversion projects such as water diversion project from the Yellow River to Baiyangdian Lake, and a water recharge project for Zhalong Wetland achieved remarkable effects.

2000年以来，塔里木河实施流域水量统一调度，保证了下游河道的水量，极大地改善了流域生态环境，胡杨林重现生机

Since 2000, unified allocation of water resources has been implemented in the Tarim River Basin, guaranteeing the water volume of downstream channels, greatly improving the ecosystem in the basin and revitalizing the Populus euphratica forest

2000年以来，黑河干流水量跨省区统一调度，使内蒙古额济纳胡杨林面积由39万亩增加到44万亩，也使周边地区的生态环境得到改善

Since 2000, unified allocation of water resources among provinces and areas has been applied to Heihe River mainstream, increasing the area of Populus euphratica forest in Ejina, Inner Mongolia Autonomous Region, from 26,000 hectares to 29,333 hectares, as well as improving the ecosystem in surrounding areas

农村水环境整治

为改善农村水生态环境，积极推进对农村河湖的全面清淤疏浚和环境整治。通过河道疏浚、岸坡整治、生态修复等工程措施，解决了河湖功能减退、水环境恶化等突出问题，构建起村村碧水相连、清水环绕的人水和谐新环境。

江苏兴化农村河道

River channels in Xinghua, Jiangsu Province

Water Environment Improvement in Rural Areas

To improve the water ecological environment in rural areas, vigorous efforts have been made to implement thorough dredging and regulation of rivers and lakes in the countryside. Structural measures such as river channel dredging, harnessing of sloping banks and ecological restoration have solved prominent problems such as the decreased functions of rivers and lakes, and the deteriorating water environment, thus creating a new environment with a harmonious human-water relationship where green and clean water is available to surrounding villages.

山西潇河河道

River channels in Xiaohe River, Shanxi Province

安徽宏村河道

River channels in Hongcun Village, Anhui Province

农村水电开发及利用

积极推进水电新农村电气化县建设、小水电代燃料生态保护建设和农村水电增效扩容改造试点。目前，全国已建成农村水电站4.5万多座，装机容量6 212万千瓦，占我国水电总装机容量的27%。电气化县户通电率达到99.8%，保护森林面积600多万亩，100多万山区农民结束了祖祖辈辈上山砍柴的历史。

Hydropower Development and Utilization in Rural Areas

To actively promote the construction of rural hydropower electrification counties, small hydropower projects have replaced wood fuel as a source of electricity, and pilot projects for the expansion of the capacity of hydropower stations have been launched in rural areas. At present, over 45,000 hydropower stations have been built in rural areas across the country with an installed capacity of 62.12 million kW, accounting for 27% of total in China. A total of 99.8% of rural counties now have access to electricity. Over 400,000 hectares of forests are protected, and more than one million farmers in mountainous regions have stopped the traditional practice of chopping firewood for fuel.

农村电力设施

Electricity facilities in rural areas

黑龙江农村电气化建设

Rural electrification, Heilongjiang Province

牧民用上了电炊具

Herdsmen can now cook with electricity

浙江开化县齐溪小水电站

Qixi small hydropower station, Kaihua County, Zhejiang Province

牧区水利建设

以发展节水灌溉饲草料地为重点，坚持“以水定草、以草定畜”，优先抓好内蒙古自治区东中部、新疆维吾尔自治区北部、青海三江源、环青海湖、四川西北部、甘肃南部等重点地区的牧区水利建设。2001—2010年发展节水灌溉饲草料地150万亩，为600多万个羊单位提供补充饲料，5 240万亩天然草场得到轮牧、休牧和禁牧。

新疆那拉提草原

Narat Grassland, Xinjiang Uygur Autonomous Region

Water Conservancy Construction in Pasture Areas

Emphasis has been placed on the development of forage land with water-saving irrigation facilities, upholding the principle of "developing forage land on the basis of water availability, and developing animal husbandry according to the capacity of forage land". Priority has been given to developing water conservancy construction in key pasture areas of middle and eastern Inner Mongolia Autonomous Region, northern Xinjiang Uygur Autonomous Region, Sanjiangyuan (joint sources of the Yangtze, Yellow and Lancang rivers) of Qinghai Province, Qinghai Lake Rim, northwestern Sichuan Province and southern Gansu Province, etc. From 2001 to 2010, 100,000 hectares of forage land with water-saving irrigation facilities were constructed, providing supplementary feed to over 6 million sheep unit, and rotational grazing and fallow grazing were implemented in 34.9 million hectares of natural grassland.

内蒙古东乌珠穆沁草原

East Ujimqin Grassland, Inner Mongolia Autonomous Region

水利风景区建设和管理

科学开发、合理利用和有效保护水利风景资源，规划建设了475个国家水利风景区和千余个省级水利风景区，形成了涵盖全国主要江河湖库、重点灌区、水土流失治理区的水利风景区群落。充分发挥水利风景区在提升工程效益、涵养水源、保护生态、改善人居环境、拉动地方经济发展等方面的作用，实现了“水清岸绿，人景相融”。

春季的上海滴水湖水利风景区

Dishuihu Scenic Water Area, Shanghai, in Spring

夏季的湖北天堂湖水利风景区

Tiantang Lake Water Scenic Area, Hubei Province, in Summer

Construction and Management of Scenic Water Areas

Scenic resources in water conservancy areas should be scientifically developed, rationally utilized and effectively protected. A total of 475 national water scenic areas and over 1,000 provincial water scenic areas have been planned and constructed, creating a community of scenic water areas covering major rivers, lakes, reservoirs, key irrigation zones, and soil and water conservation zones across the country. Scenic water areas are being brought into full play by raising project benefits, conserving water resources, protecting the ecology, improving people's living environment and stimulating local economic development. Adorned with clean water and green trees along the banks, scenic water areas exhibits a harmonious relationship between human and nature.

秋季的四川泸沽湖水利风景区

Luguhu Scenic Water Area, Sichuan Province, in Autumn

冬季的新疆天池水利风景区

Tianchi Scenic Water Area, Xinjiang Uygur Autonomous Region, in Winter

第七篇

Chapter Seven

三条红线管水

Managing Water Resources by Means of "Three Red Lines" Control

2011年中央一号文件和中央水利工作会议针对我国日益严峻的水资源形势，明确要求实行最严格水资源管理制度。2012年1月，国务院印发《关于实行最严格水资源管理制度的意见》，划定水资源开发利用控制、用水效率控制、水功能区限制纳污“三条红线”，对实行最严格水资源管理制度作出全面部署和具体安排。《意见》是指导当前和今后一个时期我国水资源管理工作的纲领性文件，在我国水资源管理史上具有里程碑意义。

With regard to the increasingly severe situation of water resources in China, the No. 1 Central Document of 2011 and the Central Working Conference on Water Resources convened in 2011 explicitly point out that the strictest water resources management system should be implemented. *The Opinion on Implementing the Strictest Water Resources Management System* issued by the State Council of China in January 2011 has made a holistic and concrete arrangement of the implementation of the system, identifying three red lines for water development and utilization control, water use efficiency control and pollutant load control in water function zones. *The Opinion* is and will be the guiding document for water resources management and is a milestone in the history of water resources management in China.

实行最严格水资源管理制度

Implementing the Strictest Water Resource Management System

三条红线
Three Red Lines

- 水资源开发利用控制红线
  Red line for water development and utilization control
- 用水效率控制红线
  Red line for water use efficiency control
- 水功能区限制纳污红线
  Red line for pollutant load control in water function zones

四项制度
Four systems

- 用水总量控制制度
  System of total water use quantity control
- 用水效率控制制度
  System of water use efficiency control
- 水功能区限制纳污制度
  System of pollutant load control in water function zones
- 水资源管理责任和考核制度
  Accountability and performance assessment system for water resources management

"三条红线"控制指标

Control Index of "Three Red Lines"

三条红线控制指标（年份） Control index of "Three Red Lines"(y)	全国用水总量（亿立方米） Water use quantity of national total ($x10^7m^3$)	万元工业增加值用水量（立方米） Industrial added value (m^3) Per 10 000 yuan RMB	农田灌溉水有效利用系数 Effective utilization coefficient of farmland irrigation water	主要江河湖泊水功能区水质达标率 Water quality compliance rate in key water function zones of rivers and lakes
2015	6 350	比2010年下降30% decrease by 30% from 2010 level	＞0.53	60%
2020	6 700	＜65*	＞0.55	80%
2030	7 000	＜40*	＞0.6	95%

* 以2000年不变价计 Calculated by the constant price of 2000

开展跨省江河流域水量分配的河流——长江流域嘉陵江

Jialing River in the Yangtze River Basin, for which water resources allocation across provincial boundaries has been carried out

严格用水总量控制

严格控制流域和区域取用水总量，制定主要江河流域水量分配方案，逐级建立取用水总量控制指标体系，实行地表水和地下水取用水总量控制。加强规划和建设项目水资源论证工作，强化地下水管理和保护。

Put Total Water Consumption Quantity Under Strict Control

The total water consumption quantity in a basin or a region should be strictly controlled. Water departments will develop water resource allocation plans for major river basins, set up a layered indicator system for total water consumption control, employ surface water and groundwater abstraction regulations, strengthen planning and water abstraction licensing mechanism for construction projects to consolidate groundwater management and protection.

已完成水量分配的大凌河

Daling River for which water resources allocation has been completed

大力推进水权制度建设，积极培育水市场，鼓励开展水权交易，运用市场机制配置水资源。全面启动全国主要跨省江河流域水量分配工作，首批25条跨省江河流域水量分配技术方案基本完成。黄河、黑河、永定河干流、大凌河等重点河流水量分配方案批复实施。宁夏回族自治区、内蒙古自治区等地区率先开展了农业和工业之间水权转换工作。

Meanwhile, we will speed up the establishment of a national water rights system and the development of the water market, encourage water rights transactions and allocate water resources by market mechanisms. Water resource allocation in major trans-provincial basins of the country has been launched and the technical schemes of water resources allocation for the first 25 trans-provincial river basins have largely been completed. Water resource allocation schemes for the Yellow River, the Heihe River, the main stream of Yongding River, Daling River and other key rivers have been approved. Regions such as Ningxia Hui Autonomous Region, Inner Mongolia Autonomous Region are taking the lead in implementing water rights transactions between the agricultural and industrial sectors.

已完成水量分配的沂河

Yihe River for which water resources allocation has been completed

严格用水效率控制

强化用水定额和计划管理，加快推进节水技术改造，加大农业、工业、服务业和城镇生活节水工作力度，加快推广普及生活节水器具，淘汰落后、耗水量高的工艺、设备和产品，积极利用再生水、雨水等非常规水源，并纳入水资源统一配置。

Strengthen Water Use Efficiency Control

Water use quota and water use planning management will be reinforced, and water-saving technology transformation will be accelerated. In addition, we will intensify efforts to save water used for agriculture, industry, service industry and urban domestic purposes, speed up the promotion of daily-use water-saving devices and eliminate outdated technology, equipment and products with high water consumption. At the same time, we are actively making use of unconventional water resources, such as recycled water and rainwater, which will be incorporated into integrated water resource allocation.

农业节水灌溉

Water-saving irrigation

利用再生水的北京奥林匹克公园中心区龙形水系

The Dragon-shaped water system in Beijing Olympic Park in which recycled water is used

2001年以来，节水型社会建设在全国范围内开展。按2000年可比价计算，全国万元工业增加值用水量从2002年的239立方米减少到2011年的113立方米，下降了52.6%。100个全国节水型社会建设试点和200个省级节水型社会建设试点工作蓬勃展开。

Actions to establish a water-saving society have been carried out across the country. Calculated by constant prices in 2000, water consumption for per 10,000 yuan of industrial added value has decreased by 52.6%, from 239 m^3 in 2002 to 113 m^3 in 2011. A total of 100 pilot projects at the national level and 200 pilot projects at the provincial level for establishing a water-saving society are in full swing.

节水型社会建设试点——甘肃张掖

Zhangye, Gansu Province, a pilot project of water-saving society develepment

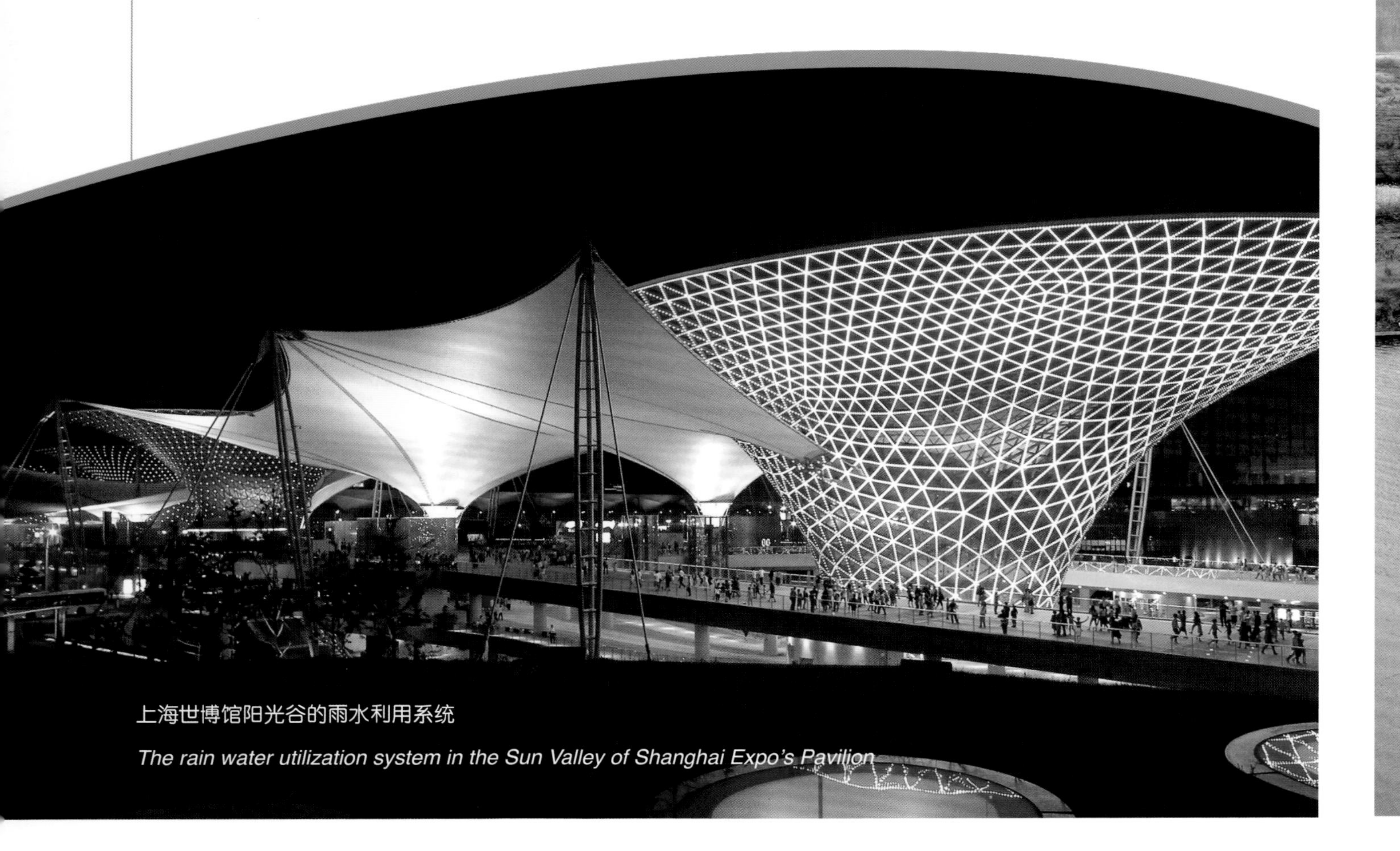

上海世博馆阳光谷的雨水利用系统

The rain water utilization system in the Sun Valley of Shanghai Expo's Pavilion

严格水功能区限制纳污

从严核定水域纳污容量，严格控制入河湖排污总量。严格水功能区监督管理，建立水功能区水质达标评价体系，加强水功能区动态监测和科学管理。加强饮用水水源地保护，依法划定饮用水水源保护区，公布重要饮用水水源地保护名录，有效应对饮用水水源地突发事件。

Restrict Pollutant Discharge in Water Function Zones

The pollutant load capacity of water bodies should be strictly defined, the total amount of pollutant discharged to rivers and lakes should be strictly controlled, and supervision and administration of water function zones should be reinforced. We will establish the water quality standard evaluation system to promote dynamic monitoring and scientific management in water function zones. We should also strengthen the protection of drinking water sources, define drinking water sources protection areas according to the law, publish the list of key drinking water source protection areas and effectively cope with emergencies in drinking water source areas.

江西婺源月亮湾

The Moon Bay in Wuyuan County, Jiangxi Province

严格落实水资源管理责任

将水资源开发、利用、节约和保护的主要指标纳入地方经济社会发展综合评价体系，县级以上地方人民政府主要负责人对本行政区域水资源管理和保护工作负总责。

Implement the Accountability System of Water Resources Management

Main indicators for development, utilization, conservation and protection of water resources will be incorporated into the local integrated performance assessment system for socioeconomic development. The local government at or above the county level will be responsible for water resource management and protection within its own jurisdiction.

浙江千岛湖

Thousand-Island Lake, Zhejiang Province

第八篇

Chapter Eight

依法科学治水

Managing Water Resources by the Law and in Scientific Way

大力推进依法治水和科技兴水，水利工作科学化水平明显提高。颁布实施了一系列水法律法规，编制实施了一大批重点水利规划，形成了完备的水法规体系和规划体系。加快水利重点领域和关键环节改革攻坚，着力构建充满活力、富有效率、更加开放、有利于科学发展的水利体制机制。加大水利科技创新步伐，水利科技总体上达到国际先进水平，一些领域处于国际领先地位。积极开展双边、多边水事交流活动，中国水利以崭新姿态登上国际舞台，国际地位和国际影响力显著提升。

We will attach great importance to the promotion of managing water resources in accordance with the law and promoting the development of the water sector with science and technology, and significantly improving the scientific level of the water sector. A series of laws and regulations related to the water sector have been promulgated and implemented, a large number of key water conservancy plans have been developed and carried out, and a comprehensive system of laws and regulations and planning in the water sector has been established. Reform in crucial areas and key links should be accelerated, and efforts should be made to develop vigorous and effective institutions and mechanisms in water sector more open and conducive to scientific development. In addition, water science and technology innovation should be advanced, and China's water science and technology on the whole has reached the international advanced level, and China is now a world leader in some fields of water science and technology. China has been actively involved in bilateral and multilateral water events, and China's water sector has stepped onto the international stage with a new face. The international status and influence of China's water sector have been improved remarkably.

水利法治建设

水法、水污染防治法、水土保持法先后修订出台，首部流域综合性行政法规《太湖流域管理条例》颁布实施，颁布了取水许可和水资源费征收管理条例、大中型水利水电工程建设征地补偿和移民安置条例、黄河水量调度条例、水文条例、抗旱条例，建立了覆盖流域、省、地、县的水行政执法网络。

Managing Water by Law

Water Law, Law on Prevention and Control of Water Pollution, and Law on Water and Soil Conservation have been amended or promulgated successively. *Management Regulations on the Taihu Lake Basin*, the first comprehensive administrative regulation for a basin, has been promulgated. Regulation on Water Abstraction Licenses and Water Resources Fee Collection and Management, Regulation on Land Acquisition Compensation and Resettlement Arrangement for Large and Medium-Sized Water Resources and Hydropower Development Project, Regulation on Water Allocation in the Yellow River Basin, Regulation on Hydrology, and Regulation on Drought Relief have been issued, and an administrative and law enforcement network on water resources covering river basins, provinces, and counties has been established.

建立了以水法为核心，包括水法、防洪法、水土保持法、水污染防治法4部法律、18件行政法规、55件部规章、800余项地方性法规和地方政府规章的水法规体系。

A water-related legal system has been established, centering on water Laws, consisiting of 4 laws, namely, water Law, Flood Control Law, Law on Water and Soil conservation and Law on the prevention and control of water pollution, 18 administrative regulations, 55 departmental statues, and over 800 local regulations.

水利规划

2012年6月，《水利发展规划（2011—2015年）》正式发布，成为指导"十二五"水利改革发展的重要依据。《全国水资源综合规划》、《全国重要江河湖泊水功能区划》、《全国抗旱规划》、《全国农村饮水安全工程"十二五"规划》、七大流域防洪规划等20多项重点水利规划得到审批，编制完成了七大流域综合规划，启动了全国水中长期供求规划和水资源保护规划编制工作，编制实施了一批水利建设专项规划。

Water Resources Planning

The Water Resources Development Plan (2011-2015) which was released in June 2012 is the key basis to guide the reform and development of the water sector during the 12th Five-Year Plan. More than 20 key water conservancy plans, including the *National Water Resources Integrated Plan, National Key Rivers and Lakes Water Function Zoning Plan, National Drought Relief Plan, National Plan of Rural Drinking Water Safety Projects During the 12th Five-Year Plan*, and flood control plans of seven major river basins have been approved. In addition, integrated plans for seven major river basins have been completed, national medium- and long-term water supply and demand planning has been launched, and a number of specific water conservancy construction plans has been formulated out and implemented.

"十二五"水利发展主要指标

Major Index of Development in the 12th Five-Year Plan

指 标 Index	2010年 The year of 2010	2015年 The year of 2015	属 性 Attribute
解决农村饮水安全人口（亿人） Rural population provided with access to safe drinking water ($\times10^8$)	[2.1]	[2.98]	约束性 Binding
水利工程新增年供水能力（亿立方米） Added annual water supply capacity by water projects ($\times10^8m^3$)	[285]	[400]	预期性 Expected
其中：新增城市年供水能力（亿立方米） Added annual urban water supply capacity ($\times10^8m^3$)	[140]	[260]	预期性 Expected
新增农田有效灌溉面积（万亩） Added effective irrigated area ($\times10^4$ ha.)	[333.33]	[266.67]	预期性 Expected
新增高效节水灌溉面积（万亩） Added high-efficiency water-saving irrigated area ($\times10^4$ ha.)	[310.67]	[333.33]	预期性 Expected
全国用水总量（亿立方米） National total water use ($\times10^8m^3$)	6 022	6 350	预期性 Expected
农田灌溉水有效利用系数 Effective coefficient of irrigated water utilization	0.50	>0.53	预期性 Expected
单位工业增加值用水量降低(%) Water consumption for per unit industrial added value decreased by (%)		[30]	预期性
新增水土流失综合治理面积（万平方公里） Controlled area of water losses and soil erosion ($\times10^4$ km^2)	[23]	[25]	预期性 Expected
新增小水电装机容量（万千瓦） Added installed capacity of small hydropower ($\times10^4$ kW)	[2 185]	[1 000]	预期性 Expected

注：[ ]内为五年累计数。
Note: numbers in "[ ]" are aggragate data for the five years

水利科技

水利科技创新体系与平台建设成效凸显，已建成2个国家重点实验室、2个国家工程技术研究中心以及10个部级重点实验室、11个部级工程技术研究中心。泥沙研究、水文水资源理论与应用研究、筑坝技术、节水技术等领域处于国际领先地位。2005—2010年，获得国家级科技奖励40项，引进、推广和转化先进适用技术800余项。

Water Science and Technology

Institutional and platform development of water science and technology innovation has achieved fruitful results. Two national key labs, two national engineering technology research centers, 10 ministerial-level key labs and 11 ministerial-level engineering technology research centers have been established. Sediment research, hydraulic and water resources theory and practice research, dam construction technology, and water-saving technology are among the most advanced in the world. Forty national science and technology prizes have been won by water institutes or agencies from 2005 to 2010, and more than 800 advanced and applicable technologies have been introduced from overseas and assimilated and disseminated in China.

长江防洪模型

The Yangtze River flood control model

黄河下游河道模型

The model of the river channel of the lower reach of the Yellow River

"节水农业技术研究与示范"获得国家科技进步二等奖

Water-saving Agriculture Technology Research and Demonstration Project, won the second prize of National Scientific and Technological Progress

浙江曹娥江大闸获"建筑工程鲁班奖"

The Cao'e River Storm Surge Dam in Zhejiang Province, a Luban Engineering Prize winner

“黄河调水调沙的理论与实践”获得2010年度国家科技进步一等奖

'Theory and Practice of Water and Sediment Regulation of the Yellow River' won the first prize of National Scientific and Technological Progress in 2010

国际合作

水利部与60多个国家和地区建立了合作关系，签署了55项双边合作协议或备忘录，建立了20多个固定交流机制。与7个周边国家和湄公河委员会建立了11个跨界河流的合作机制。中国水利专家担任国际大坝委员会、国际灌排委员会等6个国际组织主席。援助27个发展中国家进行水利水电建设，派出防洪专家组赴泰国协助抗洪，水利国际交流与合作不断拓展和深化。

International Cooperation

The Ministry of Water Resources has established cooperative relationship with more than 60 countries and regions, signed 55 bilateral cooperation agreements or MOUs and set up more than 20 regular exchange mechanisms. Besides, China has built 11 cooperative mechanisms on trans-boundary rivers with seven neighboring countries and the Mekong River Commission. Chinese experts have been elected to the presidency of six international organizations, such as the International Commission on Large Dams, and the International Commission on Irrigation and Drainage. China has provided assisstance to twenty-seven developing countries in the field of water resources and hydropower construction, and a Chinese expert team was dispatched to Thailand to assist Thailand's battle against flooding. China's international exchange and cooperation in the water sector has seen continuous expansion and deepening.

在法国马赛召开的第六届世界水论坛

The Sixth World Water Forum, Marseilles, France

泰国总理英拉访问水利部

H. E. Yingluck Shinawatra, Prime Minister of Thailand, visited the Ministry of Water Resources of China

在河南郑州召开的第五届黄河国际论坛

The Fifth International Yellow River Forum, Zhengzhou, Henan Province

在江苏南京召开的第四届长江论坛

The Fourth Yangtze River Forum, Nanjing, Jiangsu Province

在河南郑州召开的中欧水资源交流平台高层对话会

China-Europe Water Platform High-Level Dialogue Meeting, Zhengzhou, Henan Province

水利改革

初步形成了以公共财政为主导、金融支持和社会投入为补充的水利投融资新格局。建立健全了流域管理与行政区域管理相结合的水资源管理体制，全国76%的县级以上行政区实行城乡水务一体化管理。全面推行项目法人责任制、招标投标制、建设监理制，全国大中型水利工程体制改革任务基本完成。通过承包、租赁和股份合作等形式，推进小型农村水利工程产权制度改革。水利基层服务体系建设得到加强。积极推行终端水价制、超定额累进加价、丰枯季节水价、“两部制”水价等制度。水利改革发展试点全面启动。

Water Sector Reform

A new investment and financing pattern for water conservancy has been initially set up which is led by public finance and supplemented by financial support and social investment. A water resources management mechanism that combines river basin management and administrative region management has been established and improved. Integrated management of urban and rural water has been carried out in 76% of regions above county-level. A project legal person responsibility system, bidding system, and construction supervision system will be fully implemented, and large and medium-sized water engineering mechanism reform has been basically completed all over the country. Property rights reform of small-sized rural water engineering projects is being promoted by means of concessions, leasing, and joint stock partnership and so on. Grassroots service system development has witnessed further enhancement. The systems of terminal water price, escalating water price according to use volume, differentiating water price between flood and dry seasons, and "two blocks" water tariff have been promoted. Meanwhile, reform and development pilot projects in the water sector have be launched in an all-round way.

水管体制改革给水管单位带来了新的生机，图为宝鸡峡灌区渠首

Structural reform of water sector brings new vigor to water management units. The head works of Baojixia Irrigation Area, Shaanxi Province

农村河道保洁

River course cleaning in rural area

水务体制改革后城市充分利用中水浇灌绿地

Recycled water is fully used in urban areas after water system reform

北京莲石湖湿地

Lianshi Lake Wetland, Beijing

结束语

江河安澜，国泰民安，是中华民族的长久祈盼。治水兴水，造福民生，是时代赋予的神圣使命。

党的十八大进一步完善了我国新时期治水方略，深化了水利工作内涵，勾画出建设美丽中国的新蓝图。水利迎来了极好的发展机遇，处于继往开来、负重前行的关键阶段。在新的历史起点上，推进水利科学发展、和谐发展、跨越发展，必须以科学发展观为指导，认真贯彻落实党的十八大精神和2011年中央一号文件、中央水利工作会议精神，积极践行可持续发展治水思路，坚持民生优先，坚持统筹兼顾，坚持人水和谐，坚持政府主导，坚持改革创新，进一步加强水利薄弱环节建设，大力推进民生水利发展，着力夯实水利基础，加快建立水利投入稳定增长机制，全面落实最严格的水资源管理制度，不断创新发展体制机制，推动传统水利向现代水利、可持续发展水利转变，努力走出一条中国特色水利现代化道路，为全面建成小康社会、加快社会主义现代化进程提供坚实的水利支撑和保障，为实现中华民族的伟大复兴做出应有的贡献。

中华民族治水史册已经翻开了崭新的一页。让我们携起手来，关心水、爱惜水、节约水、保护水，通过我们的共同努力，实现水资源的永续利用，让水更好地造福中华民族，更多地惠泽子孙后代。

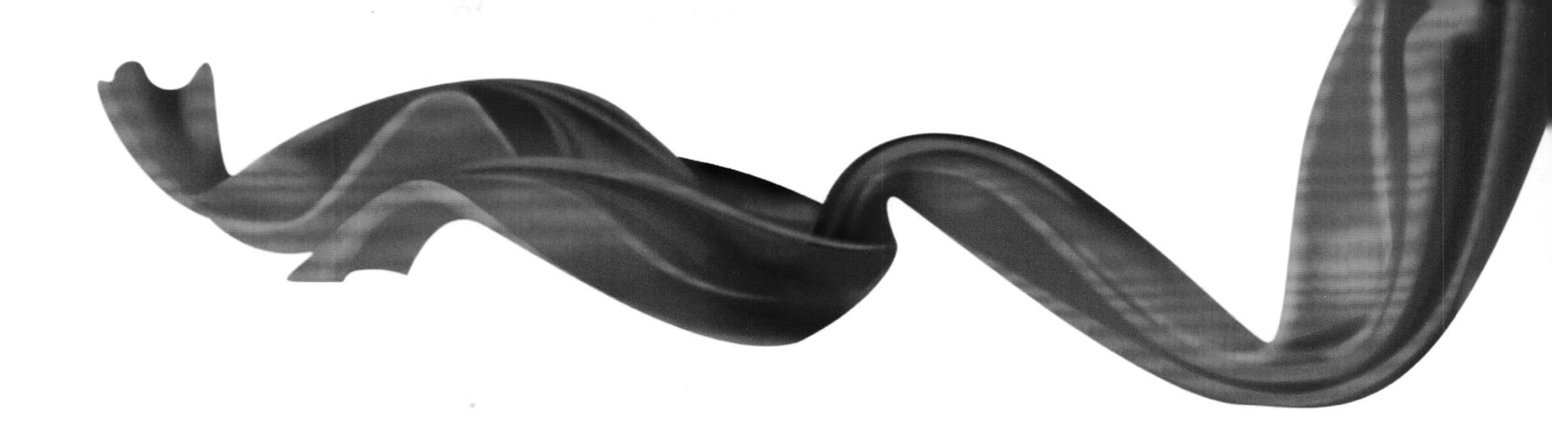

Conclusion

It is the Chinese people's everlasting expectation that all rivers run smoothly to ensure the nation's prosperity and the people's peaceful life.

It is our lofty mission bestowed by the times to manage and develop water for people's livelihood.

The 18th CPC National Congress further improved the plans and strategies of water management, enriched the content of water conservancy work and depicted a new blue print of constructing a beautiful China. The water sector in China is facing extremely good opportunities, and is at the critical point to forge ahead into the future with great obligations and responsibilities. In order to promote scientific development, harmonious development and leapfrog development for the water sector, we should stick to the Scientific Outlook on Development as our guideline, conscientiously implement the spirit of the 18th CPC National Congress and No.1 Central Document of 2011 as well as the decisions of the Central Working Conference on Water Conservancy, and actively practice sustainable development for water resources. We should stick to priorities for people's livelihood, full consideration of all factors, harmonious coexistence between people and water, the leading role of the government and reform and innovation, strengthen the weak links in the water sector, vigorously promote the development of water conservancy for people's livelihood, consolidate the basis of the water sector, and accelerate the process of establishing a stable mechanism of water investment growth. We will comprehensively implement the most stringent water resource management system, continuously innovate in water management system and mechanism, advance the transformation from traditional mode of water mangement to modern and sustainable mode of water mangement, and create a modern development pattern with Chinese characteristics for water sector. Water management will provide support and safeguard to the process of building a moderately prosperous society in all aspects, accelerate social modernization, and make contribution to the great rejuvenation of the Chinese nation.

Water management in China has now turned over a new leaf. Let us join hands to care about water, to treasure water, to save water and to protect water. Through our joint efforts, we will realize sustainable utilization of water resources and enable water to bring more benefits to the Chinese nation and our future generations.

西藏林芝巴松措

The Basum Lake in Linzhi, Tibet Autonomous Region

青海年保玉则文错湖

The Wencuo Lake in Nyainbo Yuze, Qinghai Province

编纂委员会

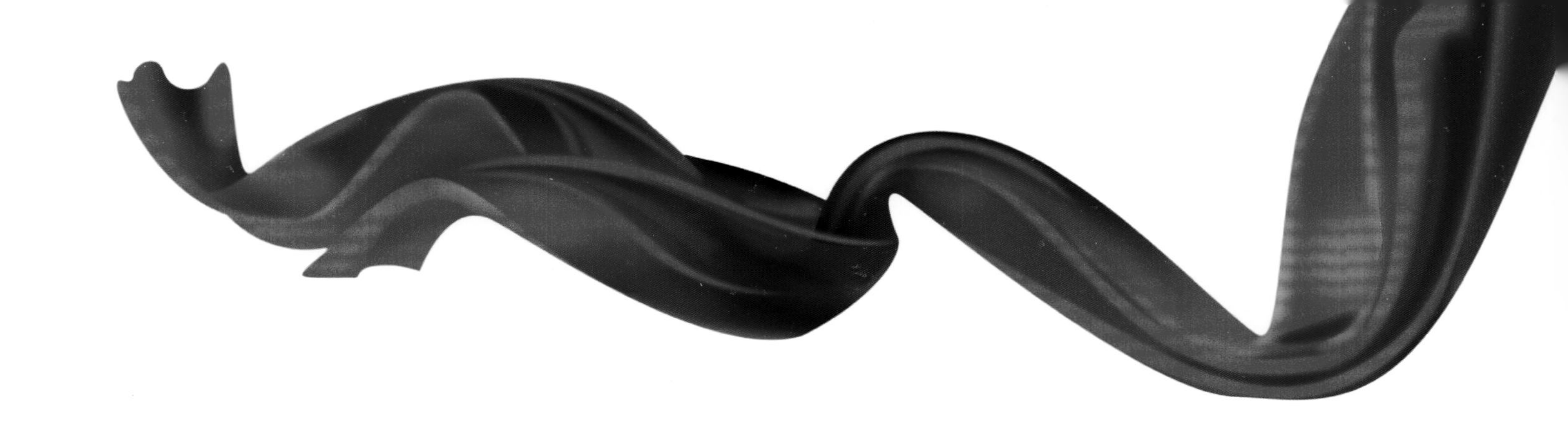

英文翻译

金　海　徐　静　沈可君　张　潭　常　远

池欣阳　彭竞君　郑磊磊　郝　钊　吴浓娣

英文校译

刘志广　金　海　梁　超

图片拍摄

（按姓氏笔画排序）

王水林　冯凯文　毕鹏飞　刘一槃　刘柏良

李先明　吴忠贤　汪　栋　张　岩　张　辉

张进平　张爱忠　陈　楠　姜拥军　殷鹤仙

翁　强　高　波　高立洪　黄正平　黄宝林

董保华　温永明　缪宜江　魏建国　等

责任编辑

徐丽娟

装帧设计

刘一槃　芦　博

图书在版编目（CIP）数据

水利 : 造福民生的伟大事业 : 汉文、英文 / 中华人民共和国水利部编. -- 北京 : 中国水利水电出版社, 2013.1
ISBN 978-7-5170-0475-2

Ⅰ. ①水… Ⅱ. ①中… Ⅲ. ①水利行业－发展－中国－画册 Ⅳ. ①TV-12

中国版本图书馆CIP数据核字(2012)第307045号

审图号：GS（2012）1487 号

书　　名	水利　造福民生的伟大事业 **Water Conservancy** A Great Cause for People's Livelihood in China
作　　者	中华人民共和国水利部　编 Compiled by the Ministry of Water Resources, P. R. China
出版发行	中国水利水电出版社　（北京市海淀区玉渊潭南路1号D座　100038） 网　址：www.waterpub.com.cn E-mail：sales@waterpub.com.cn 电　话：（010）68367658（发行部）
经　　售	北京科水图书销售中心（零售） 电　话：（010）88383994、63202643、68545874 全国各地新华书店和相关出版物销售网点
排　　版	中国水利水电出版社装帧出版部
印　　刷	北京华联印刷有限公司
规　　格	255 mm × 355 mm　8 开本　21.5 印张　80 千字
版　　次	2013年1月第1版　2013年1月第1次印刷
印　　数	0001—3500册
定　　价	380.00 元